Siti Salwa Alias · Ahmad Azmin Mohamad

Synthesis of Zinc Oxide by Sol–Gel Method for Photoelectrochemical Cells

 Springer

Siti Salwa Alias
Ahmad Azmin Mohamad
School of Materials and Mineral
   Resources Engineering
Universiti Sains Malaysia
Nibong Tebal
Malaysia

ISSN 2192-1091   ISSN 2192-1105 (electronic)
ISBN 978-981-4560-76-4 ISBN 978-981-4560-77-1 (eBook)
DOI 10.1007/978-981-4560-77-1
Springer Singapore Heidelberg New York Dordrecht London

Library of Congress Control Number: 2013953264

Printed on acid-free paper

Springer is part of Springer Science+Business Media (www.springer.com)

Preface

Various methods have been formulated and applied to produce nanocrystalline zinc oxide (ZnO) powders that are used as an electrode in a photoelectrochemical cell (PEC). The techniques in synthesizing ZnO powders, such as hydrothermal, spray pyrolysis, and laser ablation, require high temperature. This high temperature results in the chemical interactions between the material and the container walls as well as in high processing cost. In addition, the controlled precipitation in the aqueous solution used in the ZnO powder preparation tends to form zinc hydroxide rather than ZnO.

The sol–gel method can be used to synthesize ZnO powder and minimize the aforementioned limitations. This process has been more frequently used compared with other processes because it can synthesize ZnO powder at low temperatures. The sol–gel process transforms the sol or solution into a gel and then converts it into the desired final product. It is well known that the sol–gel process has been used in producing ultrafine powders, high-surface area porous materials, thin solid films, dense abrasive minerals, and continuous ceramic and glass fibers, among others.

Meanwhile, ZnO can be synthesized from diverse types of raw materials, which can affect the characteristics of the final product. Alkoxides, normally used as a precursor in the sol–gel process, are expensive and explosive materials. The metal salt zinc acetate dihydrate $[Zn(CH_3COO)_2 \cdot H_2O]$ is used as an alternative to alkoxides because its hydrolysis is easier to control.

No systematic explanations can be provided when the characteristics of ZnO synthesized by different procedures were compared with the results from the sol–gel process. In this book, ZnO was explained by sol–gel technique at various pH values, sol–gel centrifugation, and storage at room temperature. This book also focuses on the study of synthesized ZnO powder using $Zn(CH_3COO)_2 \cdot 2H_2O$ precursor, methanol (as solvent), and sodium hydroxide (NaOH) to vary the pH. The synthesized ZnO powder from the sol–gel centrifugation and sol–gel storage methods was discussed in terms of structural by using X-ray diffraction, morphological by field emission scanning electron microscopy and transmission electron microscopy, chemical interaction by Fourier-transform infrared spectroscopy, and optical properties by using UV–visible spectroscopy and photoluminescence analysis to compare the properties of the nanoparticles.

The best characteristic of the ZnO powder from centrifugation and storage of sol–gel was applied to fabricate a PEC. The current density–voltage performances of both PECs were investigated under luminescent and dark conditions.

Chapter 1 discusses the fundamentals of ZnO and the sol–gel process. Chapter 2 clarifies the effect of pH on the sol–gel method, the synthesis of the ZnO powder, and the characterization analysis. Chapter 3 discusses the effect of centrifugation and storage on the sol–gel process and the characterization analysis, which is quite important in comparing the various synthesis methods of the ZnO powder in the nanoscale level. Chapter 4 presents the application of the two types of synthesized ZnO powder as working electrodes for a PEC.

Acknowledgments

The authors would like to thank the School of Materials and Minerals Resources Engineering, Universiti Sains Malaysia, and its staff for providing good research facilities and valuable scientific knowledge. We also thank all members of the Battery Research Group for their support and valuable scientific discussions, especially to Li Jian Khoo and Ann Ling Tan for their contribution to the experiments. We would also like to thank Exploratory Research Grant Scheme, ERGS (203/PBAHAN/6730006) for the financial support of this work.

Siti Salwa Alias
Ahmad Azmin Mohamad

Contents

Chapter 1
ZnO Nanocrystalline Metal Oxide Semiconductor Via Sol Gel Method

Abstract This chapter discussed the fundamentals, properties and applications. The ZnO properties will mainly focus on structural, electronic and optical. Finally, this chapter discussed the synthesis sol–gel method.

Keywords Nanocrystalline · ZnO · Wurtzite hexagonal · Direct band gap · Sol–gel

1.1 Introduction

Nanocrystalline materials currently receive much attention because of their novel processing method, properties, and products. Metal oxide nanoparticle semiconductors, such as zinc oxide (ZnO), titanium dioxide (TiO$_2$), and zirconium oxide (ZrO), have been considered as scientific discovery for various applications [1]. These materials gained much interest because of their well-known performance in electronics, optics, photonics, and unique quantum applications, as well as their ability to alter the properties of solid state devices, which is dependent on the application [2, 3].

An early study on the nanocrystalline TiO$_2$-working electrode photoelectro-chemical cell (PEC) was developed by Grätzel et al. [4]. The large surface area associated with the use of a porous nanoparticle electrode is an important factor that influenced PEC performance. This characteristic is significant in obtaining a large volume concentration of organic dye absorption and in achieving a highly efficient PEC sunlight harvesting [5].

Generally, the properties of TiO$_2$ and ZnO oxide semiconductor materials, such as the crystal structure, point of zero charge, energy level of the conduction band (CB), and electron conductivity, influence the performance of the PEC working electrode. ZnO is apparently more flexible than TiO$_2$ as an electrode because of the wurtzite structure of ZnO compared with the anatase structure of TiO$_2$ [6]. In addition, both materials have similar band gap energies ($\sim$3.2 eV).

S. S. Alias and A. A. Mohamad, *Synthesis of Zinc Oxide by Sol–Gel Method for Photoelectrochemical Cells*, SpringerBriefs in Materials, DOI: 10.1007/978-981-4560-77-1_1, © The Author(s) 2014

ZnO has been extensively investigated because of its unique properties, which are approximately similar to those of TiO_2, although TiO_2 is the most frequently used at the beginning of the semiconductor oxide discovery. Furthermore, ZnO is an inorganic compound and exists as a mineral zincite in the earth crust. However, most commercially used ZnO is produced synthetically [7]. ZnO appears as a white powder and almost insoluble in water but soluble in any alcohol solvent, such as ethanol and methanol (CH_3OH). The ZnO powder is widely used as an additive to various materials and products such as plastics, ceramics, glass, cement, rubber, and paints [8].

1.2 ZnO: Properties and Applications

Nanostructured ZnO materials have received great attention, especially in electronics, optics, and a wide variety of photonic applications, because of their novel and fascinating optical, mechanical, electrical, thermo–electrical, and chemical properties as well as their potential technology applications. Nanomaterials based on ZnO are promising candidates for nanoelectronics and photonics [9]. In this chapter, ZnO, as well as its synthesis techniques, will be discussed, particularly the sol–gel method.

ZnO has a wurtzite hexagonal structure with lattice parameters $a = 0.325$ nm and $c = 0.521$ nm. The ZnO structure can be simply described as a number of alternating planes composed of tetrahedrally coordinated O^{2-} and Zn^{2+} ions stacked alternately along the c-axis [9]. Wang et al. [9] stated that the tetrahedral coordination of ZnO results in noncentral symmetric structure, piezoelectricity, and pyroelectricity. ZnO has polar surfaces, and the most common polar surface is the basal plane. The oppositely charged ions produce positively charged Zn-(0001) and negatively charged O-(000$\bar{1}$) surfaces, resulting in a normal dipole moment and spontaneous polarization along the c-axis as well as in the divergence in surface energy.

The atomically flat, stable, and absence of ZnO reconstruction differentiated the ZnO $\pm$ (0001) from polar surfaces that generally have facets or exhibit massive surface reconstructions to maintain a stable structure. The other two most commonly observed facets of ZnO were $\{2\bar{1}\bar{1}0\}$ and $\{01\bar{1}0\}$, which were non-polar surfaces having lower energy than the $\{0001\}$ facets [9].

The electrical conductivity of ZnO is determined by the defects in the intrinsically present material. Given that ZnO is intrinsically doped via oxygen vacancies and/or zinc interstitials, it functions as an n-type donor with activation energy between 30 and 60 meV [10].

Another important ZnO property is its optical band gap energy (E_g). ZnO is a direct band gap semiconductor with an E_g of 3.2 eV at room temperature. For relatively large ZnO particles (from 10 to 150 nm), which are substantially larger than the Bohr radius, the optical properties and density of states of the particles and the single-crystalline phase are very similar.

In ZnO, an energy gap, called band gap, exists between the two bands. The lower and upper bands are called the valence (VB) and conduction bands conduction (CB) bands, respectively. Both bands are important elements in a solar cell. All energy levels in the VB are occupied by electrons, and those in the CB are empty at a temperature of 0 K.

The band edges of semiconductor nanoparticles are known to shift to the type and amount of cations present at the interface [6]. According to Soga [11], the electrons of an isolated atom have discrete energy levels. When atoms gather to form crystals, the energy levels split separately. However, because of atomic interaction, the closely spaced levels result in a continuous energy band.

Several bonds are broken by thermal vibrations at room temperature when the band gap ranges from 0.5 to 3.0 eV. The thermal vibration will result in the creation of electrons and holes in the CB and VB, respectively. The potential energies of the CB and VB of the direct band gap ZnO semiconductor are illustrated in Fig. 1.1.

CB and VB are revealed as the top of the VB and bottom of the CB. The kinetic energy of the electron is measured upward from the CB, whereas the hole is measured downward from the VB, because a hole has a charge opposite to that of the electron. The electrons in the CB and holes in the VB contribute to the current flow. Table 1.1 summarizes the ZnO properties.

Considering that ZnO has several unique properties, it has been used in a variety of applications such as in light emitting diodes [12], lasers [13], piezo-electric transducers, varistors [14], photocatalytic degradation of environmental contaminants [15], solar cells [16], and chemical sensors [17].

Fig. 1.1 CB and VB of ZnO

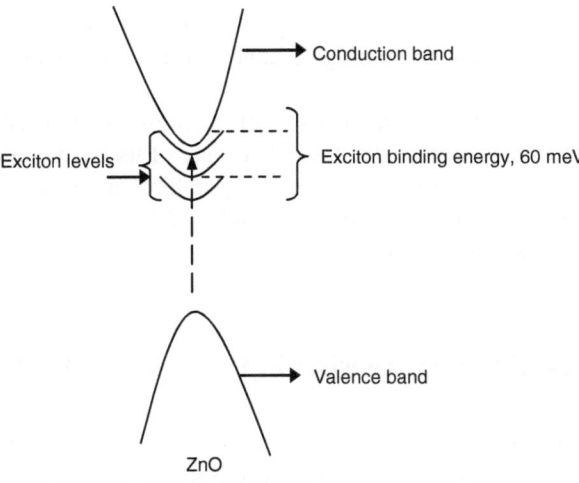

Table 1.1 ZnO properties at room temperature [6]

Properties	Characteristic
Crystal structure	Wurtize
Lattice constant (a)	0.325 nm
Lattice constant, (c)	0.521 nm
Density (kg/m^3)	5.6 gcm^{-3}
Exciton binding energy	60 meV
Static dielectric constant (ε_s)	7.9
Optical dielectric constant (ε_∞)	3.7
Optical band gap energy (E_g)	3.2 eV
Flat band potential (E_{fb})	−0.5 V versus saturated calomel electrodes (SCE)
Effective electron mass (M)	0.24–0.3 m_e, $m_e = 9.11 \times 10^{-31}$ kg
Effective hole mass (m_h)	0.45–0.6 m_e $m_e = 9.11 \times 10^{-31}$ kg
Electron mobility (μ_e)	200 cm^2 V^{-1} s^{-1}
Point of zero charge (Pzc)	8–9 pH

1.3 ZnO: Nanoparticle Synthesis Techniques

The synthesis of the one-dimensional nanostructure semiconductor oxide ZnO has been generally recently reviewed using various techniques. These techniques include the commercial method [18], sol–gel [5, 19, 20], electrodeposition [21, 22], sonochemical method [23], hydrothermal method [24], zinc–air system [25], and pyrolysis [26].

Every technique has its own advantage and can produce various ZnO nanostructures. The ZnO nanostructures can appear as simple nanoparticles, nanorods, nanobelts, branched nanorods, nanowires, ultranarrow nanobelts, hierarchical nanostructures, nanocombs, nanosprings, nanospirals, nanorings, core–shell nanostructures, nanocages, nanoflowers, and nanotubes, among other structures [27, 28].

The sol–gel process is one of the most widely used methods in the synthesis of ZnO nanoparticles [5, 19, 20] because of its good homogeneity, ease in controlling composition, low processing temperature, large-area coating, low equipment cost, and good optical properties. Table 1.2 lists the examples of ZnO synthesis techniques and structures.

1.4 ZnO: Synthesis Via the Sol–Gel Method

Four types of materials are used in synthesizing ZnO using the sol–gel method, namely, precursor, solvent, catalyst, and stabilizer. The precursor is the main material that allows the chemical reaction to occur in the sol–gel process [29].

Although precursor materials normally consist of metal ions, other elements surrounded by various reactive species called ligands are sometimes used. Metal

Table 1.2 ZnO synthesis techniques and the structures

Synthesis method	Type of structure	References
Sol–gel	Nanoparticles	[9]
Sol–gel	Nanoparticles	[10]
Sol–gel	Nanoparticles	[6]
Commercial method	Nanoparticles	[11]
Electrodeposition	Nanotubes	[12]
Sonochemical method	Nanotubes	[13]
Hydrothermal	Nanorod	[14]
Electrodeposition	Nanowires	[15]
Zinc-air system	Nanoneedles	[25]
Pyrolysis	Nanosheets	[17]

alkoxides and alkoxysilanes are most preferred because they react readily with water. A mutual solvent, such as alcohol (ethanol, CH_3OH, and isopropanol) is used because alkoxides and water are immiscible [30].

A catalyst is needed to ensure the occurrence of hydrolysis and condensation. Some examples of catalyst are hydrochloric acid, potassium hydroxide, NaOH, and ammonia. Stabilizers, such as diethanolamine, tetramethylammonium hydroxide, and ethylenediaminetetraacetate, are added to stabilize the sol. This is important to ensure the sol does not change rapidly to gel.

The properties of a particular sol–gel network are related to a number of factors that affect the rate of hydrolysis and condensation reaction, such as pH, temperature, time of reaction, reagent concentration, nature and concentration of the catalyst, aging temperature, time, and drying process.

The sol–gel process generally involves four stages, namely, hydrolysis, condensation and polymerization of monomers for particle formation, growth of particles, and agglomeration of particles. These stages are followed by the formation of networks that extend throughout the liquid medium, which results in the formation and thickening of the gel [20].

During the sol–gel process, the metal oxide particle nucleation process occurs normally by precipitation. This process involves the reaction between the soluble metal salt with the hydroxide ions (YOHs) or water. The nucleation reaction of a divalent metal salt (MX_2) and a solution that contains YOH is described as follows [31]:

$$MX_2 + 2YOH \rightarrow MO\ (s) + 2Y^+ + 2X^- + H_2O \qquad (1.1)$$

where M is the metal type (particularly Zn), X is the anion (such as $CH_3CO_2^-$, Br^-, and ClO_4^-), and Y is the cation (such as Na, Li, and K).

The growth process then occurs in the supersaturated solution until the saturation concentration of the solid is attained. After nucleation and growth, the average particle size and size distribution can be changed by aging. The two dominant aging processes are aggregation and coarsening, which depends strongly on the experimental parameters. These factors affect the final particle size and size distribution [20].

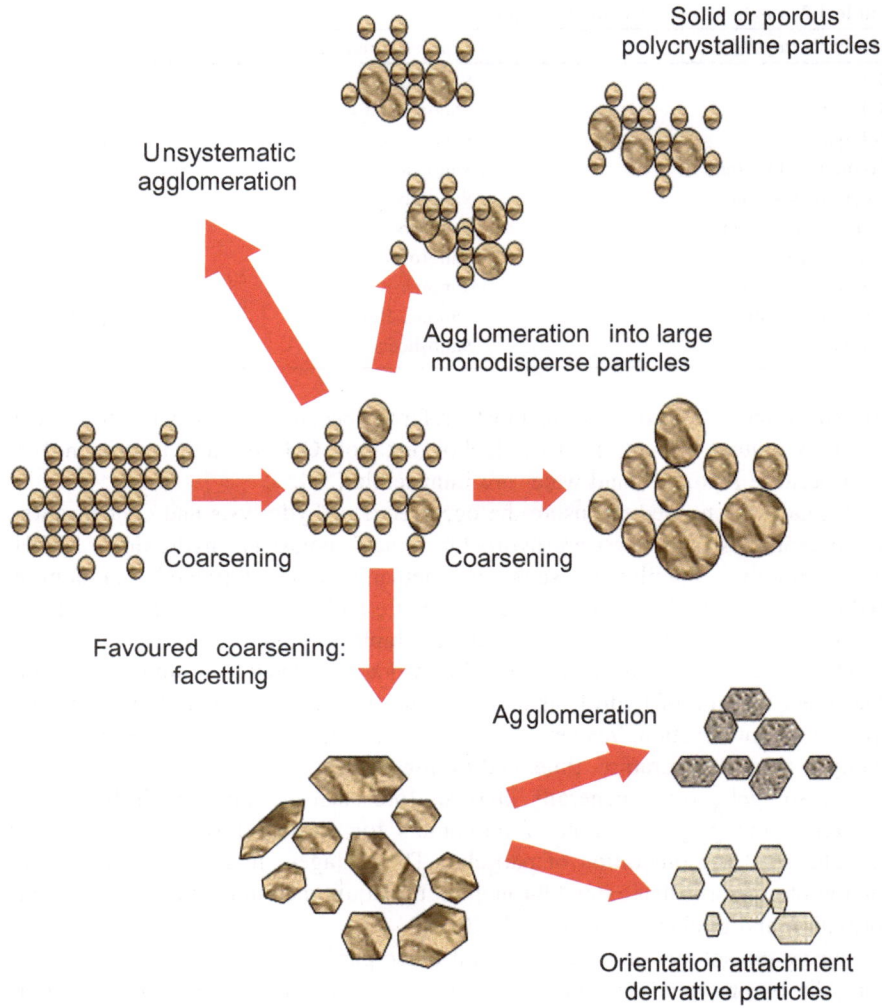

Fig. 1.2 Schematic illustration of the nanoparticle synthesis

Numerous processes indicate that nanoparticles can grow as spheres or as faceted particles, which are also affected by the aggregation process. The formation of monodispersed larger particles has been discussed extensively in colloid studies in various cases. The particles consist of a large number of nanoparticles that have aggregated, which form either compact or porous polycrystalline micrometer-sized particles.

Several routes lead to changes in the average particle size, particle morphology, and aggregation processes. A schematic illustration of the aging processes in the synthesis of nanoparticles by sol–gel process is shown in Fig. 1.2.

1.5 Summary

ZnO nanoparticles have numerous attractive structural, morphological, and optical properties for diverse applications in electrochemical device. Various techniques to synthesize ZnO are available, of which the sol–gel process is one of the most widely used methods. Four stages are involved in the sol–gel process, namely, hydrolysis, condensation and polymerization of monomers for particle formation, growth of particles, and agglomeration of particles. Numerous routes in the sol–gel process lead to changes in the average particle size, particle morphology, and aggregation processes.

References

1. Karunakaran, C., Anilkumar, P.: Photooxidation of iodide ion on immobilized semiconductor powders. Sol. Energy Mater. Sol. Cells **92**, 490–494 (2008)
2. Wang, Y., Herron, N.: Nanometer-sized semiconductor clusters: materials synthesis, quantum size effects, and photophysical properties. J. Phys. Chem. **95**, 525–532 (1991)
3. Alivisatos, A.P.: Semiconductor clusters, nanocrystals, and quantum dots. Science **271**, 933–937 (1996)
4. Gratzel, M.: Photoelectrochemical cells. Nature **414**, 338–344 (2001)
5. Meulenkamp, E.A.: Size dependence of the dissolution of ZnO nanoparticles. J. Phys. Chem. B **102**, 7764–7769 (1998)
6. Boschloo, G., Edvinsson, T., Hagfeldt, A.: Chapter 8—Dye-Sensitized Nanostructured ZnO Electrodes for Solar Cell Applications. In: Tetsuo, S. (ed.) Nanostructured Materials for Solar Energy Conversion, pp. 227–254. Elsevier, Amsterdam (2006)
7. Ozgur, U., Alivov, Y.I., Liu, C., Teke,A., Reshchikov, M.A., Dogan, S.,Avrutin, V., Cho, S.J., Morkoc, H.: Acomprehensive review of ZnO materialsand devices. J. Appl. Phys. **98**, 041301/1–041301/103 (2005)
8. Battez, A.H., González, R., Viesca, J.L., Fernández, J.E., Díaz Fernández, J.M., Machado, A., Chou, R., Riba, J.: CuO, ZrO$_2$ and ZnO nanoparticles as antiwear additive in oil lubricants. Wear **265**, 422–428 (2008)
9. Wang, Z.L.: Zinc oxide nanostructures: Growth, properties and applications. J. Phys. Condes. Matter **16**, R829–R858 (2004)
10. Look, D.C., Farlow, G.C., Reunchan, P., Limpijumnong, S., Zhang, S.B., Nordlund, K.: Evidence for native-defect donors in n-type ZnO. Phys. Rev. Lett. **95**, 225502 (2005)
11. Soga, T.: Fundamentals of Solar Cell. In: Tetsuo, S. (ed.) Nanostructured Materials for Solar Energy Conversion, pp. 3–43. Elsevier, Amsterdam (2006)
12. Kim, H., Piqué, A., Horwitz, J.S., Murata, H., Kafafi, Z.H., Gilmore, C.M., Chrisey, D.B.: Effect of aluminum doping on zinc oxide thin films grown by pulsed laser deposition for organic light-emitting devices. Thin Solid Films **377–378**, 798–802 (2000)
13. Znaidi, L., Soler Illia, G.J.A.A., Benyahia, S., Sanchez, C., Kanaev, A.V.: Oriented ZnO thin films synthesis by sol–gel process for laser application. Thin Solid Films **428**, 257–262 (2003)
14. Look, D.C.: Recent advances in ZnO materials and devices. Mater. Sci. Eng. B-Adv. Funct. Solid-State M. **80**, 383–387 (2001)
15. Sakthivel, S., Neppolian, B., Shankar, M.V., Arabindoo, B., Palanichamy, M., Murugesan, V.: Solar photocatalytic degradation of azo dye: comparison of photocatalytic efficiency of ZnO and TiO$_2$. Sol. Energy Mater. Sol. Cells **77**, 65–82 (2003)

16. Kakiuchi, K., Hosono, E., Fujihara, S.: Enhanced photoelectrochemical performance of ZnO electrodes sensitized with N-719. J. Photochem. Photobiol. A-Chem. **179**, 81–86 (2006)
17. Umar, A., Rahman, M.M., Kim, S.H., Hahn, Y.B.: Zinc oxide nanonail based chemical sensor for hydrazine detection. Chem. Commun. 166–168 (2008)
18. Saito, M., Fujihara, S.: Large photocurrent generation in dye-sensitized ZnO solar cells. Energy Environ. Sci. **1**, 280–283 (2008)
19. Hu, Z., Oskam, G., Penn, R.L., Pesika, N., Searson, P.C.: The influence of anion on the coarsening kinetics of ZnO nanoparticles. J. Phys. Chem. B **107**, 3124–3130 (2003)
20. Oskam, G.: Metal oxide nanoparticles: synthesis, characterization and application. J. Sol-Gel. Sci. Technol. **37**, 161–164 (2006)
21. Leprince-Wang, Y., Yacoubi-Ouslim, A., Wang, G.Y.: Structure study of electrodeposited ZnO nanowires. Microelectron. J. **36**, 625–628 (2005)
22. Tang, Y., Luo, L., Chen, Z., Jiang, Y., Li, B., Jia, Z., Xu, L.: Electrodeposition of ZnO nanotube arrays on TCO glass substrates. Electrochem. Commun. **9**, 289–292 (2007)
23. Chen, Y.-J., Zhu, C.-L., Xiao, G.: Ethanol sensing characteristics of ambient temperature sonochemically synthesized ZnO nanotubes. Sens. Actuator B-Chem. **129**, 639–642 (2008)
24. Guo, M., Diao, P., Wang, X., Cai, S.: The effect of hydrothermal growth temperature on preparation and photoelectrochemical performance of ZnO nanorod array films. J. Solid State Chem. **178**, 3210–3215 (2005)
25. Yap, C.K., Tan, W.C., Alias, S.S., Mohamad, A.A.: Synthesis of zinc oxide by zinc–air system. J. Alloy. Compd. **484**, 934–938 (2009)
26. Hosono, E., Fujihara, S., Honma, I., Zhou, H.: The fabrication of an upright-standing zinc oxide nanosheet for use in dye-sensitized solar cells Adv. Mater. **17**, 2091–2094 (2005)
27. Schmidt-Mende, L., MacManus-Driscoll, J.L.: ZnO—nanostructures, defects, and devices. Mater. Today **10**, 40–48 (2007)
28. Wang, Z.L.: Nanostructures of zinc oxide. Mater. Today **7**, 26–33 (2004)
29. Hench, L.L., West, J.K.: The sol-gel process. Chem. Rev. **90**, 33–72 (1990)
30. Wilson, M.: Nanopowders and nanomaterials. Nanotechnology: Basic science emerging technology 62–67 (2002)
31. Wong, E.M., Hoertz, P.G., Liang, C.J., Shi, B.-M., Meyer, G.J., Searson, P.C.:Influence of organic capping ligands on the growth kinetics of ZnO nanoparticles. Langmuir **17**, 8362–8367 (2001)

Chapter 2
ZnO: Effect of pH on the Sol–Gel Process

Abstract The technique of synthesis ZnO via sol–gel process at various pH values is presented in this chapter. The characterization methodologies such as X-ray diffraction (XRD), field-emission scanning electron microscopy (FESEM), transmission electron microscopy (TEM), crystallite and average particle size analysis, Fourier transform infrared spectroscopy (FTIR), mechanism of growth analysis, and photoluminescence (PL) analysis of ZnO are explained in details.

Keywords pH values · XRD · FESEM · TEM · FTIR · Growth analysis · PL

2.1 ZnO: Sol–Gel Process at Various pH Values

The sol pH significantly influences the ZnO properties when ZnO is synthesized by the sol–gel method. The hydrolysis and condensation behavior of the solution during gel formation are controlled by pH; thus, the morphology of ZnO is also affected [1]. Li et al. [2] showed that solution conditions particularly affect the size of the ZnO powder particle. The pH can also change the number of ZnO nuclei and growth units [3]. Sagar et al. [4] claimed that increased pH (from acidic to alkaline) of sols results in the growth of ZnO films.

Several previous studies on ZnO working electrode have focused on the synthesis of a hexagonal wurtzite-structured ZnO working electrode, with porous and nanoparticle morphology properties of ZnO in the range of ~ 50 nm, using sol–gel centrifugation from pH 6–11 [5, 6]. These studies used $Zn(CH_3COO)_2 \cdot 2H_2O$ as precursor, CH_3OH as solvent, and NaOH to adjust the pH.

Alias et al. [5] prepared ZnO solutions by adding $Zn(CH_3COO)_2 \cdot 2H_2O$ to CH_3OH at room temperature. The solutions were then stirred until a clear solution was obtained.

The clear solutions transformed into milky white slurry solutions after titration with NaOH because of the change in the pH levels of the sols from pH 6 (acidic)

S. S. Alias and A. A. Mohamad, *Synthesis of Zinc Oxide by Sol–Gel Method for Photoelectrochemical Cells*, SpringerBriefs in Materials, DOI: 10.1007/978-981-4560-77-1_2, © The Author(s) 2014

to 11 (alkaline). The resulting milky white solutions were stirred again, and all samples were left at room temperature for the sol–gel process.

The solutions were centrifuged to complete gelation and hydrolysis after 1 week. Centrifugation was performed to enhance the isolation between ZnO and the solvent. The wet, white ZnO precipitate obtained after centrifugation was dried in an oven. Finally, the dried precipitate was ground until it turned into fine white powder. The complete ZnO synthesis by the sol–gel method is depicted in Fig. 2.1.

2.2 ZnO: Structural Analysis

For structural analysis, all ZnO powders synthesized from pH 6–11 were characterized thrice using XRD (Bruker Advanced X-ray Solutions D8). The XRD results were compared with the Joint Committee on Powder Diffraction Standards (JCPDS) X-ray data file. The crystallite size of the samples was obtained by measuring the broadening of XRD peaks using the Scherrer formula plane

$$\text{Crystallite size} = \frac{180}{\pi} \cdot \frac{\kappa \cdot \lambda}{\cos \theta \cdot \sqrt{FWHM^2 - s^2}} \tag{2.1}$$

where π is equal to 3.142, λ is the wavelength of CuK_α radiation is equal to 1.5406 Å, κ is the Scherrer constant (0.89), s is the instrumental broadening (0 for Bruker Advanced X-ray Solutions D8 instrument), FWHM is the full-width at half-maximum of the (101) plane, and θ is the angle corresponding to the (101) plane. The last two values were directly obtained from *EVA* software used with XRD analysis.

The XRD profiles of ZnO powders prepared at pH 6, 7, 8, 9, 10, and 11 are demonstrated in Fig. 2.2. The peak generally changed proportional to the pH increase. No intense ZnO peak can be seen at pH 6 and 7. The two broad, intense peaks of ZnO at pH 6 and 7 indicated the existence of zinc hydroxide [$Zn(OH)_2$] in the samples. However, several high-intensity peaks at pH 8 were found at diffraction angles (2θ) of 31.68°, 34.35°, 36.16°, 47.45°, 56.48°, 62.74°, 67.84°, and 68.96°. These sets of peaks were almost similar to those ZnO synthesized from pH 9 to 11. Among the tested pH values, pH 9 had the most prominent peak with the highest intensity. The peaks slightly decreased for ZnO prepared at pH 10 and 11. The three most intense peaks of ZnO at pH 8 to 11 corresponded to the (100), (002), and (101) planes. The preferred orientation corresponding to the (101) plane was observed for ZnO at pH ≥ 8.

No intense peak was found for ZnO prepared at pH 6 and 7. Normally, the ZnO structure cannot be well synthesized at pH 6 because of the high concentration of H^+ ions and low concentration of OH^- ions in the sol. The amount of H^+ and OH^- ions are equal under a neutral condition (pH 7); therefore, both their peaks are broad. Peaks started to appear for ZnO prepared at pH 8. The peak with the highest intensity ($2\theta = 36.16°$) appeared for ZnO at pH 9 because a sufficient amount of

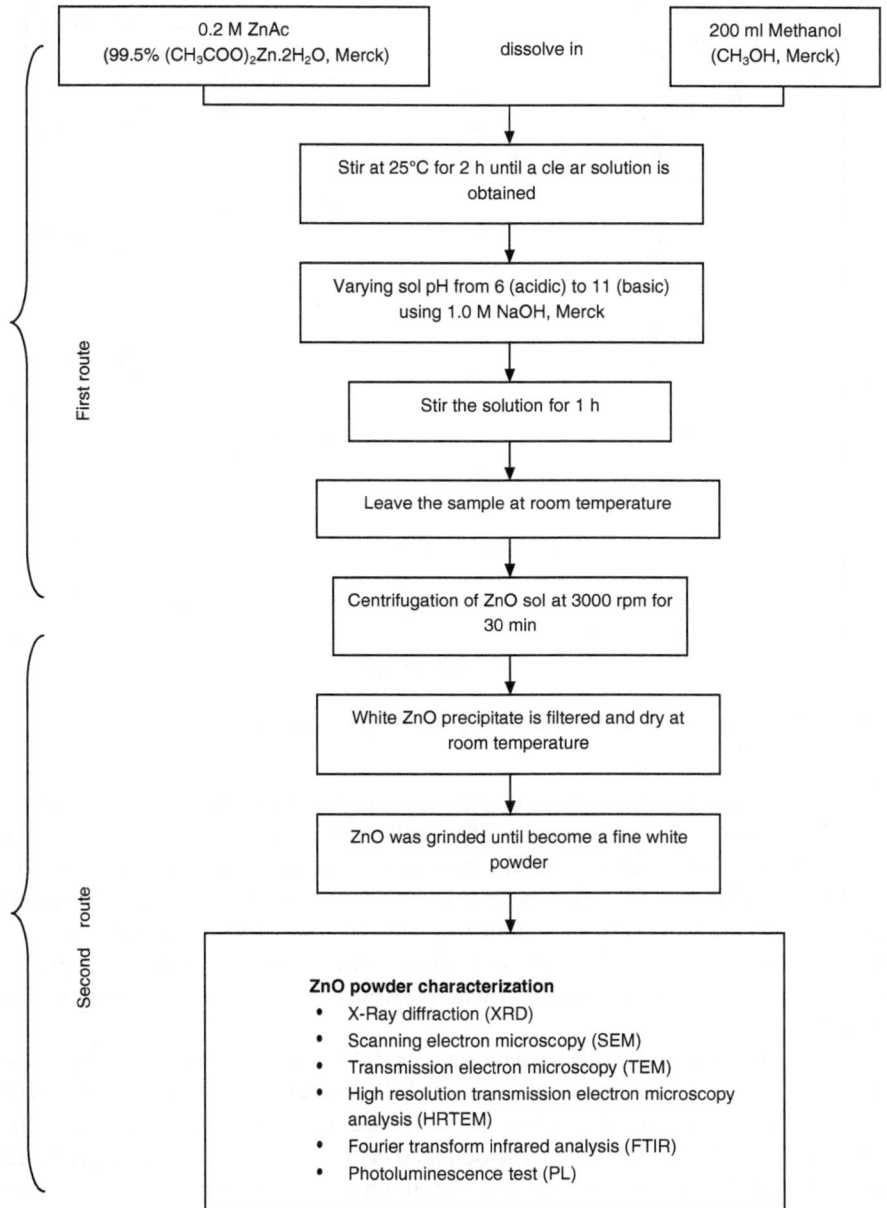

Fig. 2.1 Flow chart of experimental sol–gel methodology [7]

OH⁻ was available for ZnO formation. However, ZnO synthesized at pH 10 and 11 dissolved partly because of the high concentration of OH⁻ ions in both sols. The dissolution of large ZnO molecules produced small ZnO crystallites and caused the ZnO peak intensities to be markedly reduced.

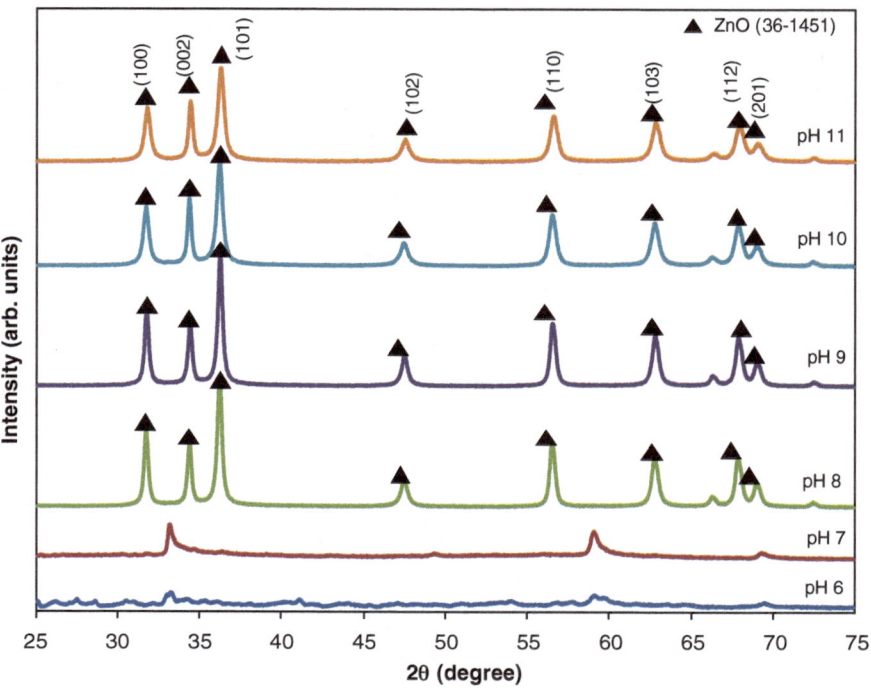

Fig. 2.2 XRD patterns of ZnO synthesized by the sol–gel process at various pH values

The observed peak intensities could be attributed to the presence of a hexagonal wurtzite structure in ZnO powder. The cell constants were $c = 5.205$ Å and $a = 3.249$ Å. These values and the wurtzite structure were confirmed by the JCPDS 36-1451 data. No other peak was detected for samples at pH ≥ 8, which indicated that all precursors completely decomposed and no other complex products formed. The XRD results confirmed that the powders synthesized in this study were ZnO, which is consistent with previously reported structural properties [8–10].

The crystallite and average particle sizes of ZnO powder as a function of pH are plotted in Fig. 2.3. The crystallite sizes of ZnO powder as a function of pH were calculated based on the (101) plane of XRD, and the average particle sizes were calculated from FESEM results. The obtained crystallite sizes of ZnO ranged from 18.37 nm to 25.36 nm. The largest (25.36 nm) and the smallest (18.37 nm) crystallite sizes were observed at pH 9 and 11, respectively. The average particle size was larger than the crystallite size.

However, the average particle size in this study ranged from 36.65 nm to 49.98 nm, which was ∼25 nm larger than the crystallite size. The average particle size analysis revealed that the largest (49.98 nm) and smallest (36.65 nm) particle sizes occurred at pH 8 and 11, respectively. The result showed that the crystallite and average particle sizes were inversely proportional to the pH values.

Fig. 2.3 Crystallite and
particle sizes of ZnO
synthesized at various pH
values

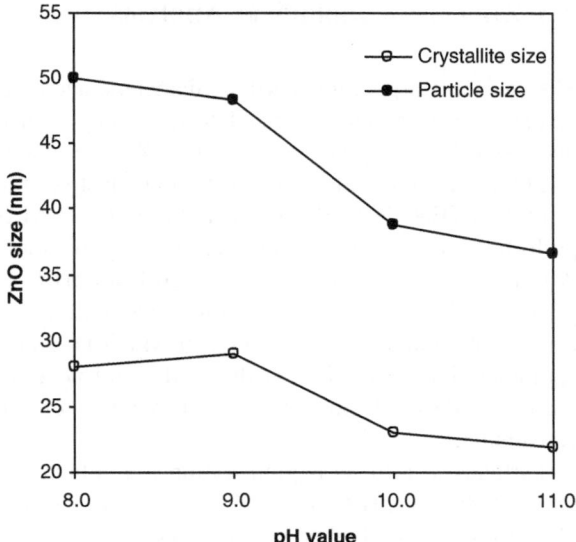

Given that ZnO peaks were absent from the XRD results, the crystallite and
particle sizes of ZnO synthesized at pH 6 and 7 could not be calculated. All ZnO
particles sizes were larger than the crystallite size, which resulted from the
abundant crystallite content of each particle. Therefore, the particle size was
equivalent to the total crystallite size inside a ZnO particle. Meanwhile, the
crystallite size corresponded to the size of one crystal inside a ZnO particle. Thus,
the crystallite size of ZnO was markedly smaller than the particle size.

The particle size decreased inversely proportional to the pH, which showed that
the amount of OH$^-$ ions played an important role in determining the ZnO particle
size. After reaching the sufficient amount of OH$^-$ ions for ZnO synthesis (pH 8
and 9), the particle size decreased because of the dissolution of ZnO. The crystallite size of ZnO increased at pH 8 and 9, which could be attributed to the
increased amount OH$^-$ ions in ZnO. However, further increase in OH$^-$ concentration in ZnO (pH 10 and 11) reduced the crystallite and particle sizes because the
amount of dissolved OH$^-$ was larger during ZnO synthesis at pH >9. ZnO dissolved when ZnO reacted with excessive OH$^-$ ions [11]. The dissolution
decreased the crystallite and particle sizes and facilitated particle agglomeration
[12]. The ZnO crystallite and particle sizes are listed in Table 2.1.

Table 2.1 Comparison
between crystallite and
particle size of ZnO
synthesized at various pH

pH	Crystallite size (nm)	Average Particle size (nm)
8	24.96	49.98
9	25.36	48.31
10	21.87	38.32
11	18.37	36.65

2.3 ZnO: Morphological Analysis

The surface morphology, particle diameters, and composition of ZnO synthesized at pH 6–11 were characterized thrice using an using FESEM and an energy-dispersive X-ray spectroscopy (EDX) (Zeiss Supra 35VP) systems at 20,000× to 100,000× magnification. Transmission electron microscopy (TEM) and high-resolution TEM (HRTEM) using a Philips TECNAI 20 series instrument were also performed at 100,000× to 1,000,000× magnification. The FESEM images of the ZnO powder nanoparticles synthesized from pH 6–11 are shown in Fig. 2.4.

A large bulk of particles with high agglomeration is shown in Fig. 2.4a, b for ZnO at pH 6 and 7, respectively, in which no single particles were found. This agglomeration resulted from the acidic and neutral pH values of $Zn(OH)_2$ sols during synthesis. Both samples lacked OH^- ions that were important for ZnO conversion.

Particles with lower agglomeration were obtained for ZnO synthesized at pH 8 than for ZnO at pH 6 and 7 (Fig. 2.4c) because of the alkaline condition in the pH 8 sample. An alkaline condition is crucial to ZnO growth [13]. The particles were homogeneous with good nanoscale structure when the pH was increased to an alkaline condition (pH 9) (Fig. 2.4d). These ZnO nanoparticles were mostly spherical.

The particles of the ZnO samples at pH 10 and 11 were small and agglomerated (Fig. 2.4e, f). Only small sections showed spherical particles. Under these highly alkaline conditions, the obtained spherical particles were smaller than the particles of ZnO synthesized at pH 9. Generally, all ZnO powders obtained in this study were nanoparticles with low agglomeration. The slight agglomeration could be attributed to the use of a centrifuge instead of the traditional sol–gel method, which did not include centrifugation [6]. Centrifugation can remove impurities such as $Zn(OH)_2$ by enhancing the interaction between Zn–O and solvent.

The OH^- ions allow the nucleation and growth of ZnO as well as particle formation. Particle sizes decrease when the pH of the sols is >9. Wu et al. [14] suggested that large particles consist of agglomerated nanoparticles. Either compact or porous polycrystalline microsized particles can be formed using the colloidal sol–gel technique. In this part, the particle sizes at low pH values were nanosized even when agglomeration occurred. The average particles of ZnO synthesized between pH 9 and 11 were uniform in shape and size ranging from 36.65 to 49.98 nm in diameter.

The chemical composition of the ZnO powder synthesized between pH 6 and 11 were predicted using EDX, and the results are shown in Fig. 2.5. All samples revealed high peak intensities.

The atomic percentages of Zn and O in ZnO prepared at pH 6 and 7 were lower than those of ZnO prepared at other pH values. This finding could be attributed to the insufficient OH^- ions during ZnO synthesis (Fig. 2.5a, b). However, ZnO synthesized between pH 8 and 11 showed higher atomic percentages of Zn and O because of the sufficient amount of OH^- ions during ZnO synthesis (Fig. 2.5c–f).

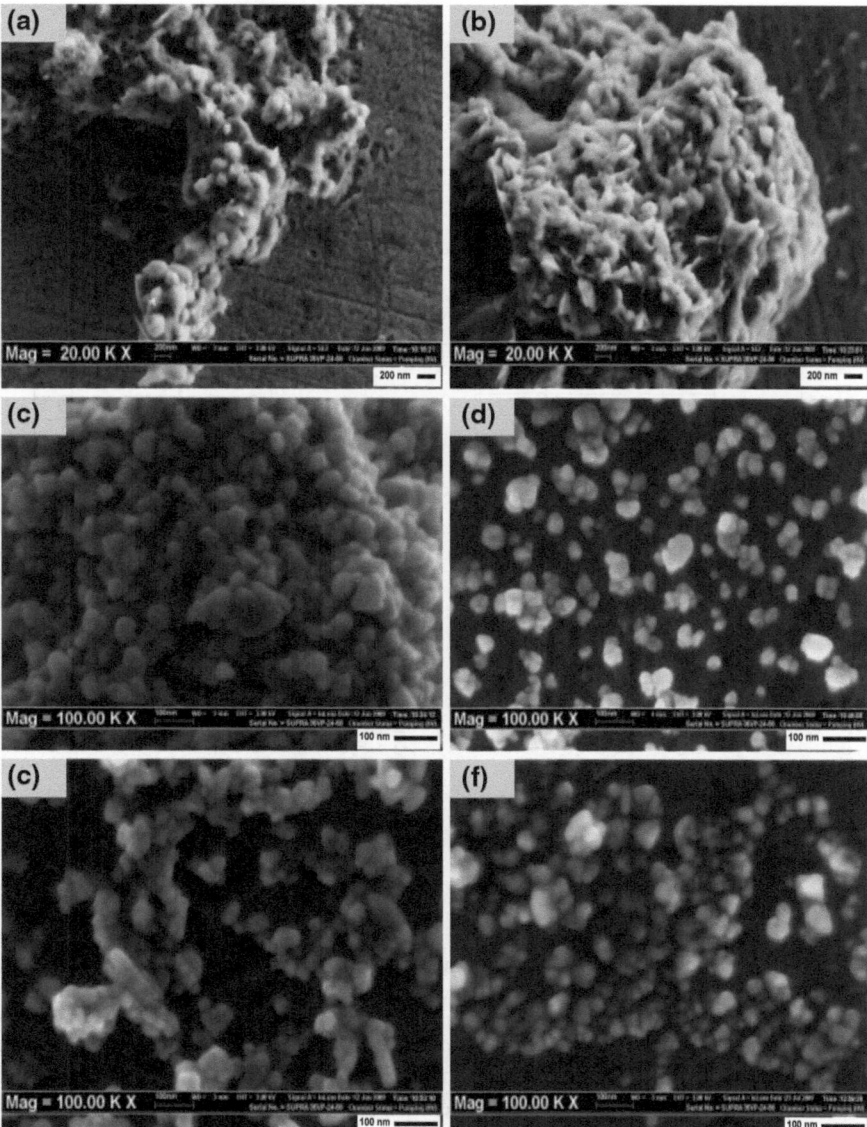

Fig. 2.4 FESEM images of ZnO powder synthesized at pH **a** 6, **b** 7, **c** 8, **d** 9, **e** 10, and **f** 11

The atomic percentages of Zn and O were generally close to the stoichiometric composition and had a ratio of approximately 1:1. This result confirmed that the powder contained only Zn and O elements and that ZnO successfully formed.

TEM was performed to study further the morphology and crystallinity of the ZnO powder nanoparticles at pH 9. The typical TEM images of ZnO nanoparticles are illustrated in Fig. 2.6. The large particles (aggregates) consisting of very small

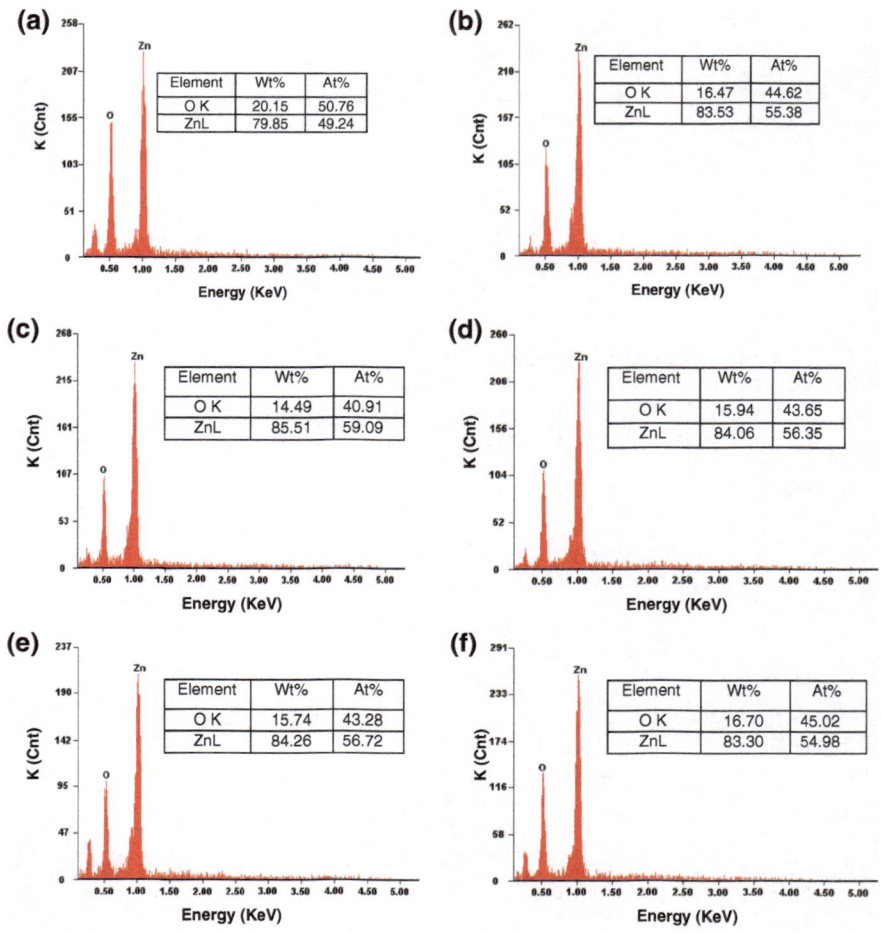

Fig. 2.5 EDX of ZnO powder synthesized at pH **a** 6, **b** 7, **c** 8, **d** 9, **e** 10, and **f** 11

particles were much clearer than those in the FESEM images. However, the particles were still uniform in terms of shape distribution even if the agglomeration of particles was low. In the TEM images, ZnO had particle sizes with diameters ranging from 26.00 to 48.10 nm. These sizes were close to the diameter of the average particles obtained by FESEM analysis (49.98 nm).

The HRTEM image of ZnO with well-resolved lattice fringe nanoparticle crystals is demonstrated in Fig. 2.7. The lattice spacing was approximately 5.900 Å between adjacent lattice planes. This finding further confirmed that the preferential growth direction of ZnO nanoparticles was the c-axis [001] direction. This growth direction was consistent with the ZnO properties (wurtzite hexagonal structure with lattice parameters $a = 0.325$ nm and $c = 0.521$ nm) [15]. The lattice spacing value was larger than the lattice constant of ZnO in the JCPDS

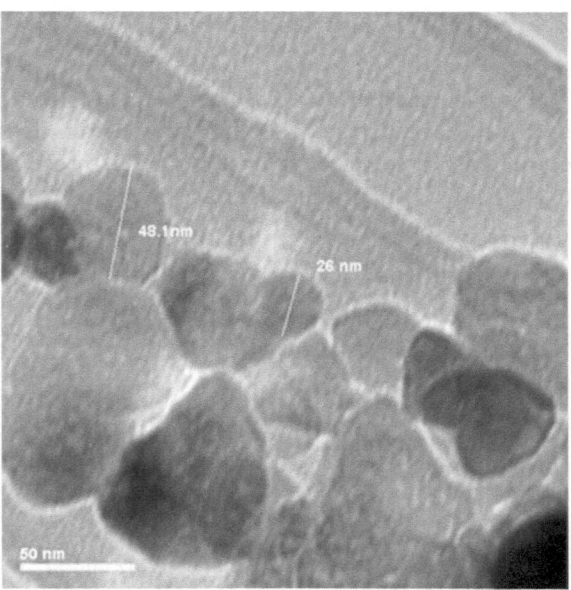

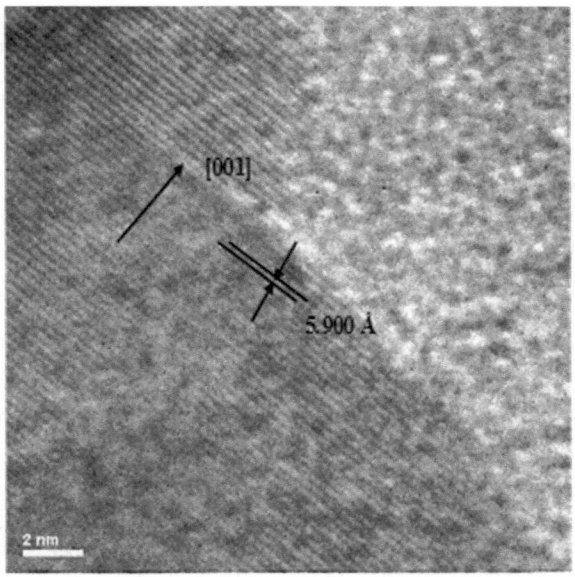

36-1451 XRD data file (5.205 Å). This discrepancy could be attributed to the shape of particles and the size of ZnO obtained from this synthesis technique. The TEM and HRTEM results were similar to the previous findings on synthesized ZnO nanocrystals [16–18].

2.4 ZnO: Growth Mechanism

The growth mechanism of ZnO nanoparticles powder during the sol–gel process as described in reactions (3.2)–(3.6) is shown in Fig. 2.8. To grow ZnO with high crystallinity, a sufficient amount of OH^- is needed to dissolve ZnO. Sol–gel formation can be divided into four stages;

(i)   Hydrolysis
(ii) Condensation and polymerization (nucleation) of monomers for particle formation
(iii) Growth of particles
(iv) Aging.

At pH ≥ 8.0, $Zn(OH)_4^{2-}$ was converted to ZnO because of the high chemical potential of OH^- at equilibrium with a dehydration reaction [19]

$$OH^- + OH^- \rightarrow O^{2-} + H_2O \tag{2.2}$$

To allow condensation to occur, zero-charge $Zn(CH_3COO)_2 \cdot 2H_2O$ precursor molecules (Fig. 2.8) were needed to form a specific amount of water or NaOH. During the initial stage of the reaction, aqueous $Zn(CH_3COO)_2 \cdot 2H_2O$ solution reacted with NaOH to form $Zn(OH)_2$, sodium acetate (CH_3COONa), and water molecules [20]. This transformation can be represented as follows [21, 22]:

$$Zn(CH_3COO)_2 \cdot 2H_2O + 2NaOH \rightarrow Zn(OH)_2 + 2CH_3COONa + 2H_2O \tag{2.3}$$

The build-up of precursor molecules resulted in solution supersaturation (Fig. 2.8b). Nucleation continuously occurred above and below the condensation and polymerization threshold. $Zn(OH)_2$ reacted with the water molecule to form the growth unit $Zn(OH)_4^{2-}$ and hydrogen ions ($2H^+$) as follows:

$$Zn(OH)_2 + 2H_2O \rightarrow Zn(OH)_4^{2-} + 2H^+ \tag{2.4}$$

Further ZnO growth and agglomerate separation from the supersaturated solution was possible until the solution became saturated with a white precipitate of colloidal-gel $Zn(OH)_4^{2-}$ (Fig. 2.8c). Centrifugation transformed $Zn(OH)_4^{2-}$ into ZnO according to the following reaction [23]:

$$Zn(OH)_4^{2-} \rightarrow ZnO + H_2O + 2OH^- \tag{2.5}$$

Oskam [20] stated that the two dominant processes during aging are aggregation and coarsening. For this synthesis technique, the aging rates of nanoparticles strongly depended on the pH. This parameter determined the structure and size distribution of the final spherical nanoparticles (Fig. 2.8d). Centrifugation eliminated the agglomeration of incompletely transformed $Zn(OH)_4^{2-}$ and spun $Zn(OH)_4^{2-}$ to enable complete transformation to ZnO.

However, high OH^- concentrations (pH 10 and 11) caused ZnO to react with OH^-. This ZnO dissolution during the reverse reaction produced $Zn(OH)_4^{2-}$ according to this reaction [11]:

Fig. 2.8 Proposed growth
mechanism of ZnO
nanoparticles

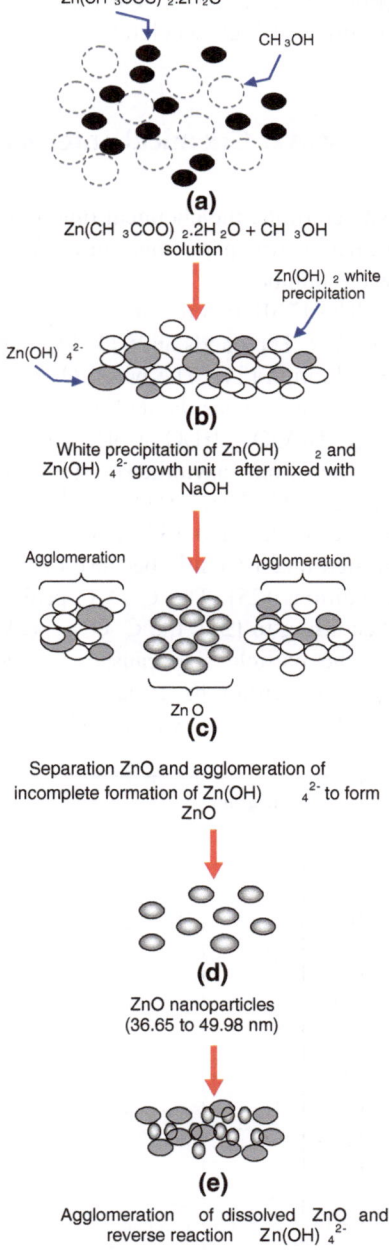

$$ZnO + H_2O + 2OH^- \rightarrow Zn(OH)_4^{2-} \qquad (2.6)$$

The large ZnO molecules then produced small ZnO particles. Consequently, the combination of these small particles with $Zn(OH)_4^{2-}$ induced agglomeration

(Fig. 2.8e). Although centrifugation reduced such agglomeration, dissolution still occurred at high acceleration.

2.5 ZnO: Chemical Interaction Analysis

To determine the chemical functional group of ZnO (pH 6–11), the FTIR (Perkin-Elmer®) was performed thrice for all ZnO samples between 4000 and 400 cm^{-1} wave number.

The FTIR spectra of ZnO synthesized at various pH values are illustrated in Fig. 2.9. The bands at 3543–3393 cm^{-1} in Region 1 correspond to the O–H mode of vibration [24]. The broad O–H peaks became narrower with an increase in the pH value because the additional amount of O–H from the NaOH reacted with the $Zn(CH_3COO)_2 \cdot 2H_2O$ at pH values ≥ 8.

The strong asymmetric stretching mode of vibration of C=O was observed between 1612 and 1560 cm^{-1} in Region 2. The symmetric stretching occurred between 1453 and 1333 cm^{-1} because of the presence of C–O. Both peaks were shifted inconsistently because of the diverging structural morphologies in alkaline conditions [25]. The C–O–C peak was observed between 1023 and 1020 cm^{-1}. According to [21], the C–O–C peak usually appears at 1256 cm^{-1}.

The particle size caused a large shift in the IR peak. The ZnO peak that appeared within the range of 550 to 453 cm^{-1} indicated that the transformation of $Zn(OH)_2$ to ZnO was completed. Previous studies have found the ZnO peak

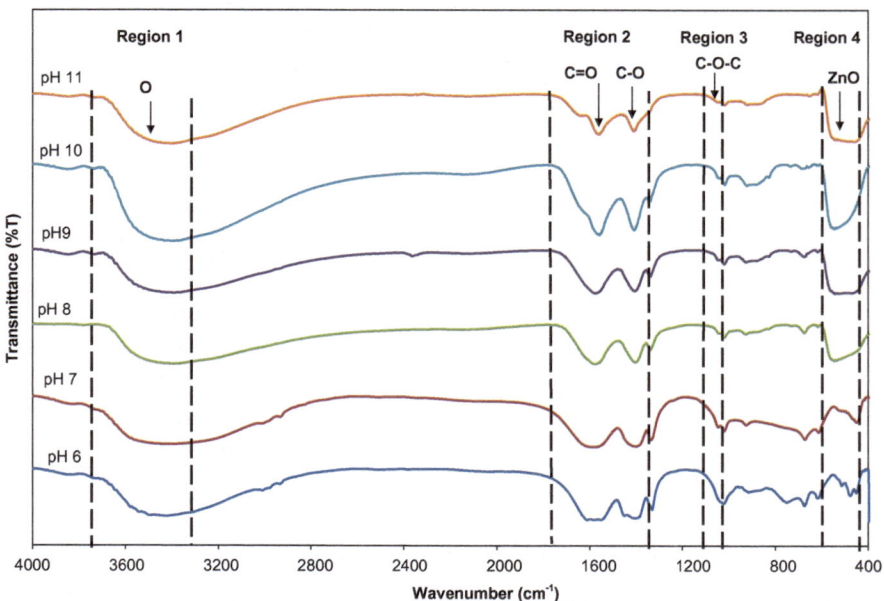

Fig. 2.9 FTIR spectra of ZnO synthesized at various pH values

Table 2.2 FTIR of ZnO at various pH values

pH	Wave number (cm^{-1})	Functional group
6	3543	O–H
	1612–1584,1453–1333	C=O, C–O
	1023–1020	C–O–C
	515–453	ZnO
7	3411	O–H
	1585, 1404–1340	C=O, C–O
	1023–1020	C–O–C
	453	ZnO
8	3393	O–H
	1579, 1407–1341	C=O, C–O
	1023–1020	C–O–C
	457	ZnO
9	3392	O–H
	1579, 1411–1341	C=O, C–O
	1023–1020	C–O–C
	543	ZnO
10	3400	O–H
	1560, 1412–1346	C=O, C–O
	1023–1020	C–O–C
	550	ZnO
11	3393	O–H
	1560, 1412	C=O, C–O
	1023–1020	C–O–C
	550	ZnO

between 464 and 419 cm^{-1} [8, 26, 27]. In this part, the ZnO peaks of all samples were shifted inconsistently because of the effect of ZnO particles which changes the particle size along with pH [28]. Thus, the average particle size affects the peak shifts. The FTIR results support the FESEM results.

The existence of other functional groups aside from ZnO and O–H such as C=O, C–O, and C–O–C from $Zn(CH_3COO)_2 \cdot 2H_2O$ precursor and CH_3OH solvent, do not have much effect on the synthesized ZnO powders. Results from the FTIR analysis of ZnO are listed in Table 2.2.

2.6 ZnO: Optical Analysis

The optical properties and E_g values of the ZnO nanoparticles (pH 8–11) were analyzed thrice using a PL spectrometer (Jobin Ynon HR 800 UV) between 300 and 600 nm wave number. The source of laser was HeCd (325 nm, 20 mW). The E_g values were calculated from the highest peak intensity (y-axis) intercept with wave number (x-axis) using

$$E_g = hv = (h \times c)/\lambda \tag{2.7}$$

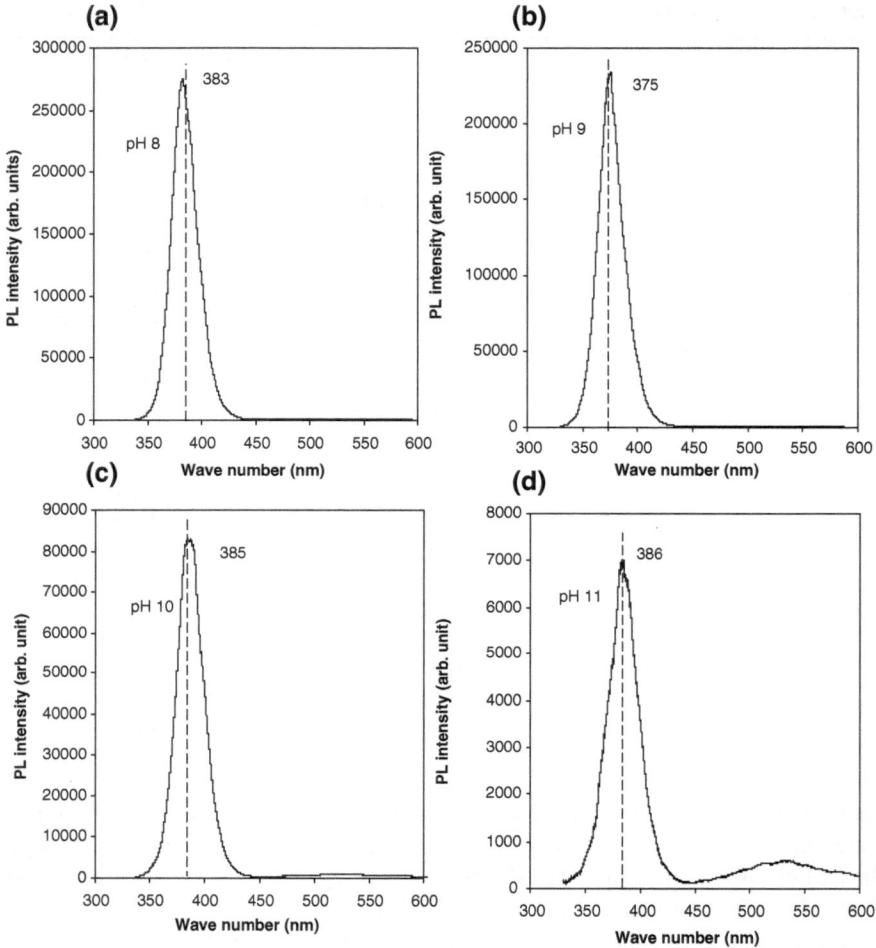

Fig. 2.10 PL analysis of ZnO powder at pH **a** 8, **b** 9, **c** 10, and **d** 11

where h is Planck's constant (6.62×10^{-34}), c is the speed of light (3×10^{8} ms^{-1}), and λ is the wave number between 300 and 600 nm.

The optical properties of the ZnO powders were analyzed using PL analysis, and the results are shown in Fig. 2.10. The ZnO synthesized at pH 8 reveals a strong UV emission band of 383 nm (Fig. 2.10a). The UV emission band of ZnO at pH 9 which was slightly decreased to 375 nm is illustrated in Fig. 2.10. When the pH values were increased to pH 10 and 11, the UV emission band increased broadly to 385 and 386 nm, respectively, as demonstrated in Fig. 2.10c, d.

The E_g of the ZnO samples at pH 8 to 11 was calculated based on the wave numbers using Eq. (2.7). The highest E_g value of 3.31 eV was observed at ZnO synthesized at pH 9. Meanwhile, E_g values for the ZnO samples at pH 8, 10, and 11 were slightly lower compared with the optimum values at 3.24, 3.22, and

pH	Energy gap, E_g (eV)
8	3.24
9	3.31
10	3.22
11	3.21

Table 2.3 Energy gap of ZnO synthesized at various pH levels from PL analysis

3.21 eV, respectively. The shift in the E_g tendency was related mainly to the confinement effect of the small size of ZnO [29]. The ZnO synthesized by this method exhibited a direct energy band gap given that the E_g ranged from 3.21 to 3.31 eV.

However, a broad intense peak was found at 510–580 nm in ZnO at pH 11. This result was due to the existence of a green emission that corresponded to the single ionized oxygen vacancy. In addition, the green emission is caused by the recombination of photogenerated hole with single ionized charge state of the vacancy. The low broad intense peak of green emission also presents the low concentration of oxygen vacancies, which was not considerably affected by the ZnO optical properties. Results from the PL analysis were almost similar to the previous work on synthesized ZnO [30–32]. The E_g values at different pH levels are shown in Table 2.3.

2.7 Summary

The ZnO nanoparticle powders were effectively synthesized using sol–gel method at pH 6–11. The ZnO synthesized at pH 9 achieved the best characteristic with the largest crystallite size (25.36 nm) based on the XRD result. At pH 9, FESEM and TEM analyses showed that the average diameters of the ZnO particle were 48.31 and 48.10 nm, respectively. The HRTEM shows that the ZnO lattice spacing was approximately 5.900 Å between adjacent lattice planes. The preferential growth direction of the ZnO nanoparticles was oriented on the c-axis [001] direction. The FTIR results showed the appearance of the 543 cm^{-1} peak, which depicted the existence of the ZnO functional group at pH 9. The average particle size also affected the FTIR shifted peaks. In the PL analysis, the ZnO at pH 9 exhibited the widest E_g enhancement of 3.31 eV.

References

1. Wahab, R., Ansari, S.G., Kim, Y.S., Song, M., Shin, H.-S.: The role of pH variation on the growth of zinc oxide nanostructures. Appl. Surf. Sci. **255**, 4891–4896 (2009)
2. Li, W.J., Shi, E.W., Fukuda, T.: Particle size of powders under hydrothermal conditions. Cryst. Res. Technol. **38**, 847–858 (2003)
3. Zhang, H., Xiangyang, Ji, Y., Xu, J., Que, D.,Yang, D.: Synthesis of flower-like ZnO nanostructures by an organic-free hydrothermal process. Nanotechnology **15**, 622–626 (2004)

4. Sagar, P., Shishodia, P.K., Mehra, R.M.: Influence of pH value on the quality of sol–gel derived ZnO films. Appl. Surf. Sci. **253**, 5419–5424 (2007)
5. Alias, S.S., Ismail, A.B., Mohamad, A.A.: Effect of pH on ZnO nanoparticle properties synthesized by sol–gel centrifugation. J. Alloy. Compd. **499**, 231–237 (2010)
6. Rani, S., Suri, P., Shishodia, P.K., Mehra, R.M.: Synthesis of nanocrystalline ZnO powder via sol–gel route for dye-sensitized solar cells. Sol. Energy Mater. Sol. Cells **92**, 1639–1645 (2008)
7. Alias, S.S.: Synthesis and characterization of agar gel polymer electrolyte and zinc oxide for photoelectrochemical cell, Masters of Science, Universiti Sains Malaysia, Nibong Tebal, Penang, Malaysia (2011)
8. Anna, K., Nina, P., Yuri, K., Meinhard, M., Werner, Z., Aharon, G.: Coating zinc oxide submicron crystals on poly(methyl methacrylate) chips and spheres via ultrasound irradiation. Ultrason. Sonochem. **15**, 839–845 (2008)
9. Maiti, U.N., Ahmed, S.F., Mitra, M.K., Chattopadhyay, K.K.: Novel low temperature synthesis of ZnO nanostructures and its efficient field emission property. Mater. Res. Bull. **44**, 134–139 (2009)
10. Mazloumi, M., Taghavi, S., Arami, H., Zanganeh, S., Kajbafvala, A., Shayegh, M.R., Sadrnezhaad, S.K.: Self-assembly of ZnO nanoparticles and subsequent formation of hollow microspheres. J. Alloy. Compd. **468**, 303–307 (2009)
11. Daneshvar, N., Aber, S., Seyed Dorraji, M.S., Khataee, A.R., Rasoulifard, M.H.: Photocatalytic degradation of the insecticide diazinon in the presence of prepared nanocrystalline ZnO powders under irradiation of UV-C light. Sep. Purif. Technol. **58**, 91–98 (2007)
12. Li, P., Liu, H., Xu, F.-x.,Wei, Y.: Controllable growth of ZnO nanowhiskers by a simple solution route. Mater. Chem. Phys. **112**, 393–397 (2008)
13. Zhang, J., Sun, Yin, Su, Liao,Yan: Control of ZnO Morphology via a simple solution route. Chem. Mat. **14**, 4172–4177 (2002)
14. Wu, L., Wu, Y., Lü, Y.: Self-assembly of small ZnO nanoparticles toward flake-like single crystals. Mater. Res. Bull. **41**, 128–133 (2006)
15. Wang, Z.L.: Zinc oxide nanostructures: growth, properties and applications. J. Phys.-Condes. Matter **16**, R829–R858 (2004)
16. Jang, J.-M., Kim, S.-D., Choi, H.-M., Kim, J.-Y., Jung, W.-G.: Morphology change of self-assembled ZnO 3D nanostructures with different pH in the simple hydrothermal process. Mater. Chem. Phys. **113**, 389–394 (2009)
17. Lin, C–.C., Li, Y–.Y.: Synthesis of ZnO nanowires by thermal decomposition of zinc acetate dihydrate. Mater. Chem. Phys. **113**, 334–337 (2009)
18. Musić, S., Dragčević, Đ., Popović, S., Ivanda, M.: Precipitation of ZnO particles and their properties. Mater. Lett. **59**, 2388–2393 (2005)
19. Hosono, E., Fujihara, S., Kimura, T., Imai, H.: Non-basic solution routes to prepare ZnO nanoparticles. J. Sol–Gel. Sci. Technol. **29**, 71–79 (2004)
20. Oskam, G.: Metal oxide nanoparticles: synthesis, characterization and application. J. Sol–Gel. Sci. Technol. **37**, 161–164 (2006)
21. Du, H., Yuan, F., Huang, S., Li, J., Zhu, Y.: A new reaction to ZnO nanoparticles. Chem. Lett. **33**, 770–771 (2004)
22. Li, W.J., Shi, E.W., Zhong, W.Z., Yin, Z.W.: Growth mechanism and growth habit of oxide crystals. J. Cryst. Growth **203**, 186–196 (1999)
23. Uekawa, N., Yamashita, R., Jun Wu, Y.,Kakegawa, K.: Effect of alkali metal hydroxide on formation processes of zinc oxide crystallites from aqueous solutions containing $Zn(OH)_4^{2-}$ ions. Phys. Chem. Chem. Phys. **6**, 442–446 (2004)
24. Fernandes, D.M., Silva, R., Hechenleitner, A.A.W., Radovanovic, E., Melo, M.A.C., Pineda, E.A.G.: Synthesis and characterization of ZnO, CuO and a mixed Zn and Cu oxide. Mater. Chem. Phys. **115**, 110–115 (2009)

25. Usui, H.: Surfactant concentration dependence of structure and photocatalytic properties of zinc oxide rods prepared using chemical synthesis in aqueous solutions. J. Coll Interf Sci. **336**, 667–674 (2009)

26. Li, R., Jiang, Z., Chen, F., Yang, H., Guan, Y.: Hydrogen bonded structure of water and aqueous solutions of sodium halides: a Raman spectroscopic study. J. Mol. Struct. **707**, 83–88 (2004)

27. Wahab, R., Ansari, S.G., Kim, Y.-S., Seo, H.-K., Shin, H.-S.: Room temperature synthesis of needle-shaped ZnO nanorods via sonochemical method. Appl. Surf. Sci. **253**, 7622–7626 (2007)

28. Wahab, R., Ansari, S.G., Kim, Y.S., Dar, M.A., Shin, H.-S.: Synthesis and characterization of hydrozincite and its conversion into zinc oxide nanoparticles. J. Alloy. Compd. **461**, 66–71 (2008)

29. Hung, C.-H., Whang, W.-T.: A novel low-temperature growth and characterization of single crystal ZnO nanorods. Mater. Chem. Phys. **82**, 705–710 (2003)

30. Fan, D.H., Zhu, Y.F., Shen, W.Z., Lu, J.J.: Synthesis and optical properties of hierarchical pure ZnO nanostructures. Mater. Res. Bull. **43**, 3433–3440 (2008)

31. Qiu, M., Ye, Z., Lu, J., He, H., Huang, J., Zhu, L., Zhao, B.: Growth and properties of ZnO nanorod and nanonails by thermal evaporation. Appl. Surf. Sci. **255**, 3972–3976 (2009)

32. Tian, Y., Lu, H.-B., Liao, L., Li, J.-C., Wu, Y., Fu, Q.: Controlled growth surface morphology of ZnO hollow microspheres by growth temperature. Physica E **41**, 729–733 (2009)

Chapter 3
ZnO: Effect of Centrifugation and Storage on Sol–Gel Process

Abstract The properties of ZnO synthesized by sol–gel centrifugation (SGC) and sol–gel storage (SGS) are presented in this chapter. Furthermore, the physical appearance of ZnO has been discussed. The ZnO structural, morphological and optical properties are characterize using X-ray diffraction (XRD), field-emission scanning electron microscopy (FESEM), transmission electron microscopy (TEM) and UV-visible (UV-vis) spectroscopy.

Keywords Sol–gel centrifugation · Sol–gel storage · XRD · FESEM · TEM · UV–vis

3.1 ZnO: Comparison Between Sol–Gel Centrifugation and Sol–Gel Storage

ZnO nanoparticles can also be synthesized using divalent metal salts and an aqueous solution. Centrifugation is frequently used to separate species of different sizes, masses, or densities. The mixture with denser components will migrate away from the axis of the centrifuge, and the less-dense components will migrate toward the axis. The Coriolis force influences crystal growth, and well-defined size characteristics can be generated by this method [1]. In fact, centrifugation [2] can improve the structural, morphological, and optical properties of ZnO compared with storage method [3].

ZnO powders were synthesized via two different procedures in the sol–gel process. Some powders were synthesized using SGS, in which the ZnO gel was stored for 48 h, and the rest were synthesized using SGC.

The alcoholic precursor solution was prepared by adding 0.2 M $(CH_3COO)_2Zn·2H_2O$ into CH_3OH that resulted in a transparent sol with no precipitate and turbidity. This alcoholic precursor solution preparation was used to transform the physical state of $(CH_3COO)_2Zn·2H_2O$ from solid to liquid. In

S. S. Alias and A. A. Mohamad, *Synthesis of Zinc Oxide by Sol–Gel Method for Photoelectrochemical Cells*, SpringerBriefs in Materials, DOI: 10.1007/978-981-4560-77-1_3, © The Author(s) 2014

addition, the intermolecular bonding in $(CH_3COO)_2Zn \cdot 2H_2O$ also weakened. These conditions facilitated the occurrence of reactions of the starting materials.

After the preparation of the alcoholic precursor solution, the beaker was covered with parafilm to prevent the exposure of fumes generated during stirring. The solution was then stirred using a magnetic stirrer to aid the dissolution of $(CH_3COO)_2Zn \cdot 2H_2O$ into CH_3OH and prevent particle agglomeration. Stirring also prevented the $(CH_3COO)_2Zn \cdot 2H_2O$ particles from precipitating at the bottom of the beaker. A clear solution of sol was prepared after stirring. The prepared sol was stable and transparent without any precipitate or turbidity.

After the alcoholic precursor solution was stirred, the pH of the solution was adjusted to pH 9 by NaOH titration. The pH level of the milky white ZnO was measured using a pH meter. NaOH was diluted from concentrated NaOH by adding concentrated NaOH white granules into distilled water in a volumetric flask. Next, at pH 9, the solution was stirred again with a magnetic stirrer to thoroughly mix and homogenize NaOH with the solution. The beaker was covered with a parafilm to prevent exposure from fumes in the environment.

In SGS, the stirred sample was stored for 48 h to allow the completion of the gelation and hydrolysis processes as well as the slow crystallization and precipitation of the white ZnO precipitates of ZnO at the bottom of the beaker. The sample was then filtered using a filter paper and washed with excess CH_3OH to remove the starting materials.

In the centrifuge procedure, a centrifuge machine by Rotina 38 was used to separate ZnO particles and the solution by centrifugal force acting onto the sample from the machine. The sample, after it was stirred for a second time, was placed into the centrifuge machine. The paste-like ZnO was collected after centrifugation by pouring out the solution from the container.

To obtain pure ZnO, the paste-like ZnO obtained from the two procedures was dried in an oven for 2 h at 120 °C and decomposed to release the organic compounds in the sample. The dried ZnO samples were ground using a mortar and pestle to obtain ZnO in powder form. A flow chart of the general ZnO powder synthesis and characterization is demonstrated in Fig. 3.1.

3.2 ZnO: Physical Appearance

The physical appearance of sample during the synthesis of the ZnO powders are illustrated in Figs. 3.2, 3.3, 3.4, 3.5 and 3.6. The transparent alcoholic precursor solution after the addition of $(CH_3COO)_2Zn \cdot 2H_2O$ to CH_3OH is depicted in Fig. 3.2. The milky white ZnO sample after the addition NaOH to the sol at pH 9 is shown in Fig. 3.3. The ZnO sample stored for 48 h to complete the gelation and hydrolysis processes is shown in Fig. 3.4. The resulting ZnO gel obtained after separating ZnO from the solution using the centrifugation or storage process is shown in Fig. 3.5. The ZnO product after drying is shown in Fig. 3.6.

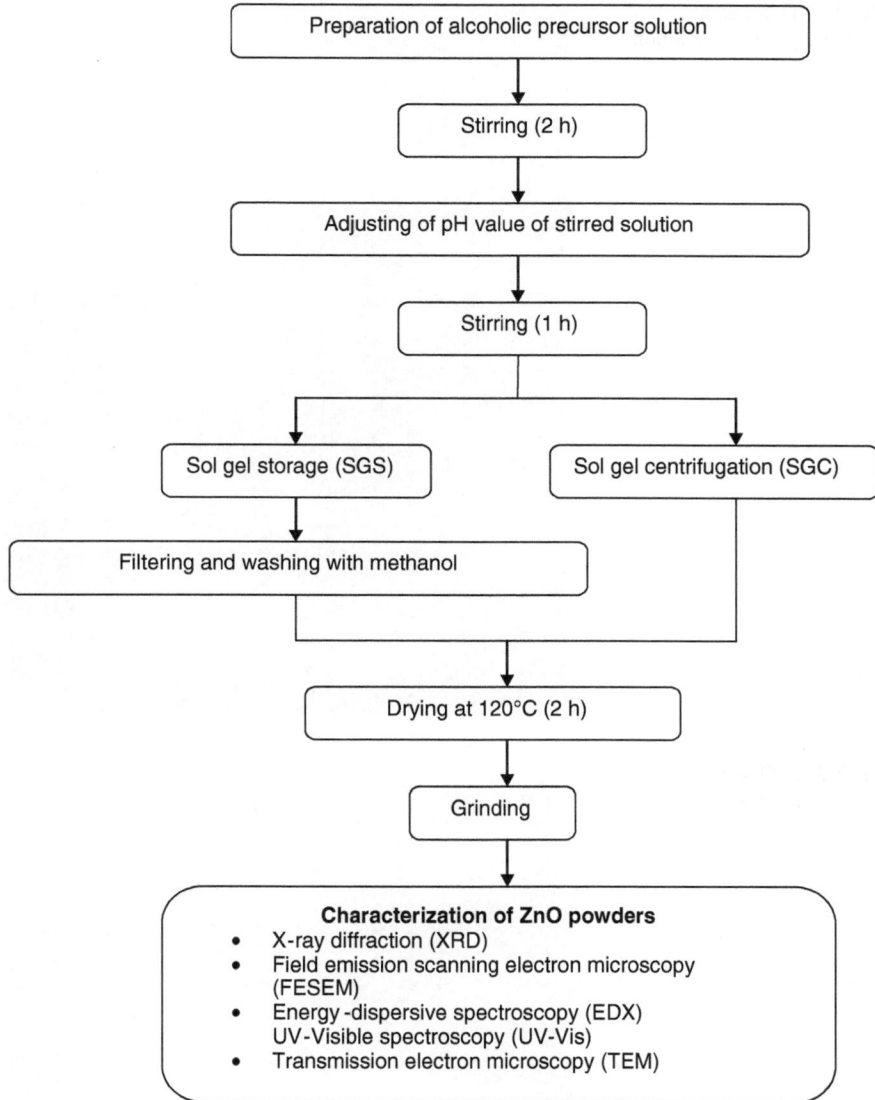

Fig. 3.1 Flow chart of the synthesis and characterization of ZnO powders using sol–gel storage and sol–gel centrifugation [4]

3.3 ZnO: Structural and Crystallite Size Analysis

The structure and composition of the synthesized ZnO powder was analyzed using Bruker Advanced X-ray Solutions D8. The crystallite size of the crystals was calculated by measuring the broadening of the diffraction lines corresponding to the maximum peak intensity and by the Scherrer formula [Eq. (2.1)].

Fig. 3.2 Transparent
alcoholic precursor solution

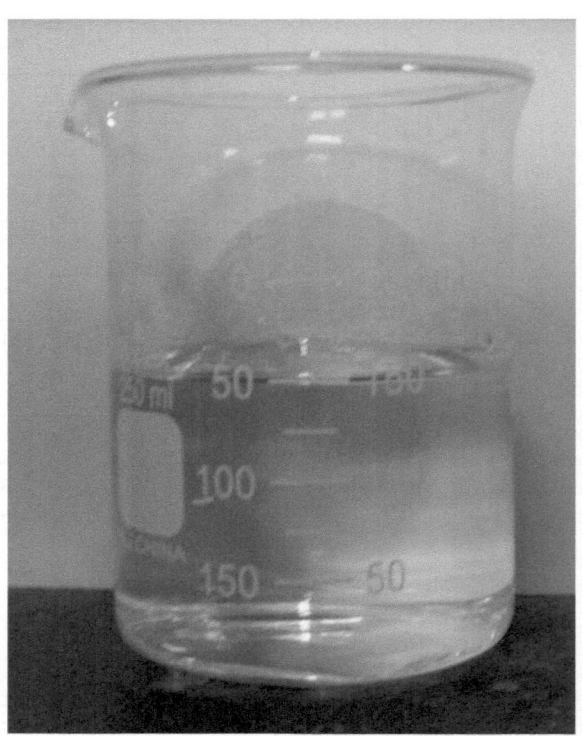

Fig. 3.3 ZnO sample with
milky white color

Fig. 3.4 ZnO sample after storage for 48 h

Fig. 3.5 Paste-like ZnO sample after the separation process

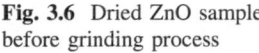
Fig. 3.6 Dried ZnO sample before grinding process

The XRD patterns for the SGS and SGC powders depicted in Fig. 3.7 show that the ZnO for both powders are crystalline in nature. The peaks of the XRD patterns for both samples corresponded to the (100), (002), (101), (102), and (110) planes, among others, and the preferred orientation is the (101) plane. The relative intensities, spacing values, and angles of diffraction (2θ) of the peaks in the XRD patterns from both samples coincided with the JCPDS Card No. 36-1451 for ZnO. All the XRD results confirmed that the synthesized powders in this study consisted of single phase hexagonal wurtzite ZnO.

Although both samples exhibited XRD patterns that indicated that they comprised ZnO, the peak intensities for both samples were different. The peak intensity for SGC powder was higher than that of the SGS powder. This result showed that the SGC powder is more crystalline or has higher crystal quality than the SGS powder [5]. The XRD patterns of the SGS and SGC powders were similar to the patterns obtained by Maensiri et al. [6] and Rani et al. [3].

Based on the Scherrer formula, the average diameters of the ZnO crystallites for the SGS and SGC powders were calculated to be 56 and 58 nm, respectively. The intensities of the SGC diffraction peaks were higher than those of the SGS, demonstrating that SGC crystallites had larger diameters than SGS. These findings are similar to those of a previous research on crystallite sizes [7, 8].

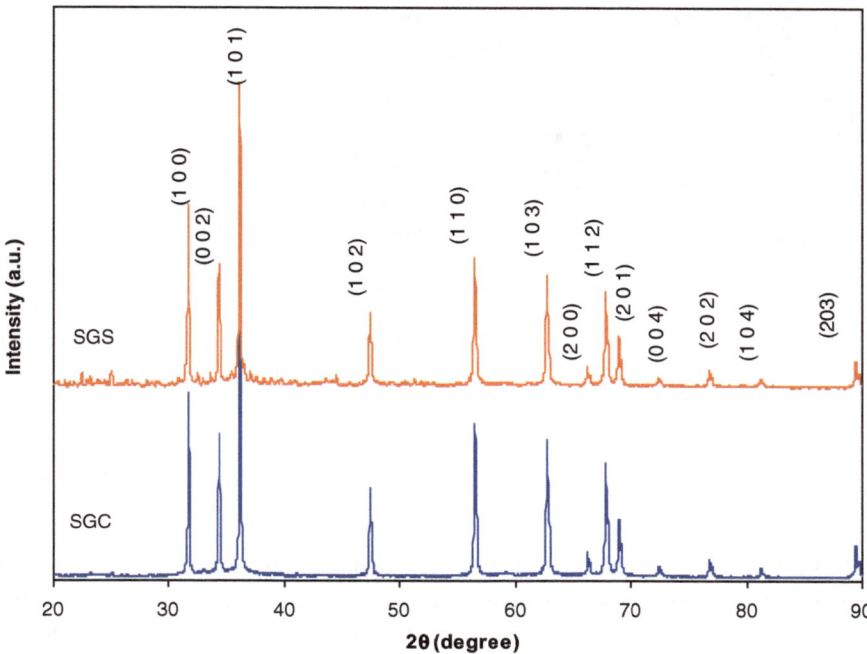

Fig. 3.7 XRD pattern of the ZnO powder synthesized using the sol–gel route for the SGS and SGC samples

3.4 ZnO: Morphological Analysis

The surface morphologies, particle diameters, and compositions of ZnO synthesized by centrifugation and storage were investigated using FESEM and EDX (Zeiss Supra 35VP). The ZnO powders were first diluted in CH_3OH and dispersed ultrasonically to break the physically linked particle clusters. One drop from each sample suspension was then placed onto different copper plates, which were ground and polished to obtain a smooth and flat surface. Then, spin coating was performed to spread and homogenize the ZnO particles on the copper plates. The CH_3OH on the copper plates was then evaporated in air. After CH_3OH evaporation, the samples were subjected to FESEM and EDX analyses.

The FESEM images of ZnO powders synthesized by SGS and SGC are illustrated in Fig. 3.8. Large and inconsistent hexagonal-shaped ZnO particles, with average size ranging from 330 to 530 nm, were observed for the SGS sample (Fig. 3.8a). However, the average particle size distribution of the SGC sample was almost uniform and ranged from 20 to 80 nm (Fig. 3.8b).

The ZnO obtained by SGC is a single-phase nanoparticle (not agglomerate), unlike the ZnO obtained by SGS. In SGC, the ZnO precipitates obtained after the sol–gel process were repeatedly washed with CH_3OH during centrifugation. Centrifugation is an effective technique for removing impurities such as $Zn(OH)_2$

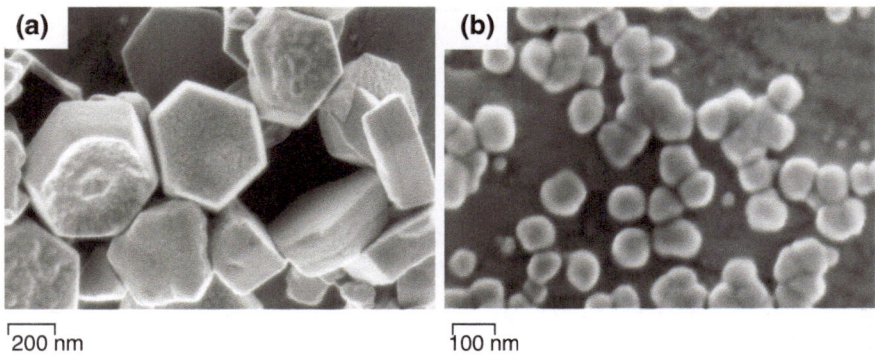

Fig. 3.8 FESEM images of ZnO powders synthesized using **a** SGS and **b** SGC

by isolating the ZnO in the CH_3OH solvent, which was previously demonstrated by Hosono et al. [9]. Without centrifugation, the particles grew in size and continued to mix with the impurities.

Results from the FESEM analysis showed that the particle sizes in both samples were larger than the crystallite sizes determined from the XRD patterns. The observed particles consisted of single primary nanocrystallites and secondary ZnO particles formed by the fusion of several primary crystallites. Thus, ZnO primary crystallite aggregation constructed the larger polycrystalline particles observed in FESEM [3].

The chemical compositions of the synthesized ZnO powders were assessed with EDX (Fig. 3.9). The relative atomic fractions of Zn and O were close to the stoichiometric composition of ZnO in both SGS and SGC samples (Fig. 3.9a, b) with a ratio of approximately 1:1. This analysis also confirmed that the powders contained only Zn and O elements.

The ZnO nanoparticles were further analyzed by TEM using Philips CM12 equipped with an analysis Docu Version 3.2 image analysis system. The samples

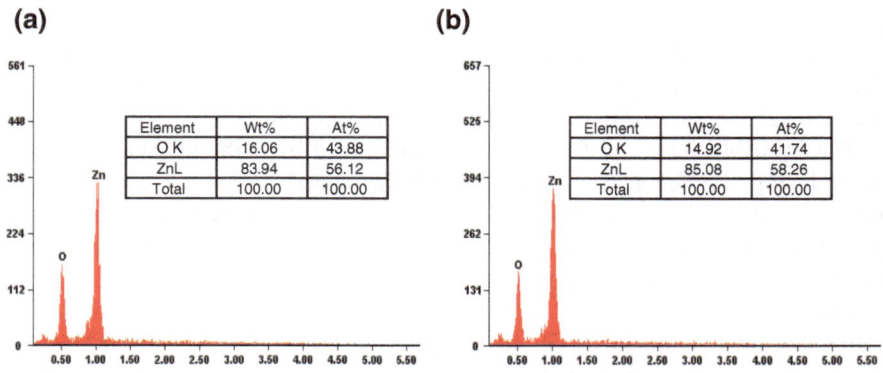

Fig. 3.9 EDX of ZnO powders synthesized by **a** SGS and **b** SGC

were initially prepared by suspending the ZnO powder in CH_3OH, followed by ultrasonic dispersion. A droplet of the resulting suspension was placed on a carbon film coated with 400 mesh copper grid and was given 1 to 3 min to react. After the large particles have settled to the bottom of the flask, a drop was transferred into a pipette. The droplet was wicked to dry using pieces of filter paper. The grid was then placed in a filter paper-lined petri dish until it can be examined.

The TEM images of the nanocrystalline SGC ZnO powder illustrated in Fig. 3.10 demonstrate that the ZnO particles were hexagonal and not agglomerated. The average particle size was determined to be 55 nm, which is comparable to the particle size range found by FESEM (20 to 80 nm). Moreover, hexagonally shaped crystallites may not be stable, as indicated by FESEM analysis. Throughout FE-SEM analysis, the surfaces tended to evolve, probably because the surface energies could be reduced to a more stable form. The TEM image results are similar in shape and size to previous studies on synthesized ZnO nanocrystals [10–12].

Most of the particles were between 45 and 65 nm, with a normal size distribution (Fig. 3.11). This result suggests that the size of most of the ZnO particles in SGC is close to the average particle size. In addition, the TEM images obtained for SGC show that the shape of the ZnO particles are similar to those reported by Rani et al. [3] and Elkhidir Suliman et al. [13]. These studies produced hexagonal ZnO nanoparticles.

Fig. 3.10 TEM image of ZnO SGC powder

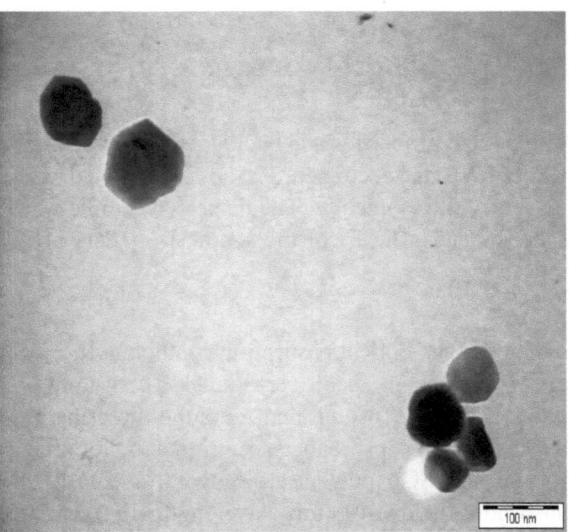

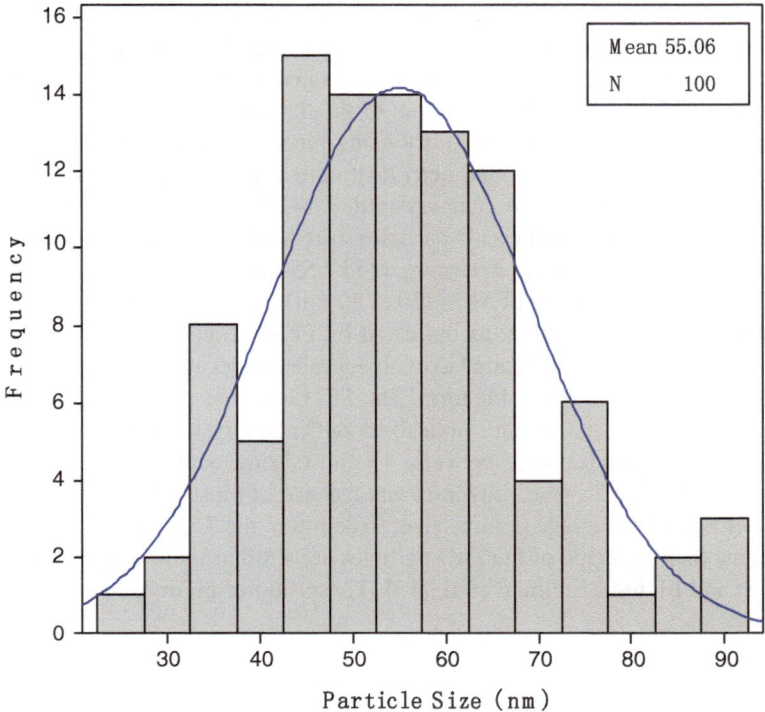

Fig. 3.11 Histogram of the particle size for 100 particles in SGC

3.5 ZnO: Optical Analysis

The optical properties and band gap of ZnO powder were analyzed and determined using UV–vis spectroscopy. The ZnO powder was first suspended in CH_3OH and dispersed ultrasonically. The dispersed sample was then examined using a UV–vis spectrometer. The E_g of the synthesized ZnO powder was determined as

$$(\alpha h v)^2 = A\left(h v - E_g\right) \tag{3.1}$$

where α is the optical absorption coefficient [$\alpha = (1/d)\ln(1/T)$, with d = thickness and T = transmittance], h is Planck's constant, v is the frequency of the incident photon, A is a constant, and E_g is the direct band gap.

The UV–vis absorbance spectrum and plot of absorption as a function of energy for the ZnO SGC powder are shown in Fig. 3.12. The ZnO powder showed strong absorption below 400 nm wavelength (Fig. 3.12a), which is suitable for the fabrication of photochemical cells because it can absorb UV radiation from the sun. The sharp decrease in the intensity of transmitted light at approximately 400 nm was due to band edge absorption [14].

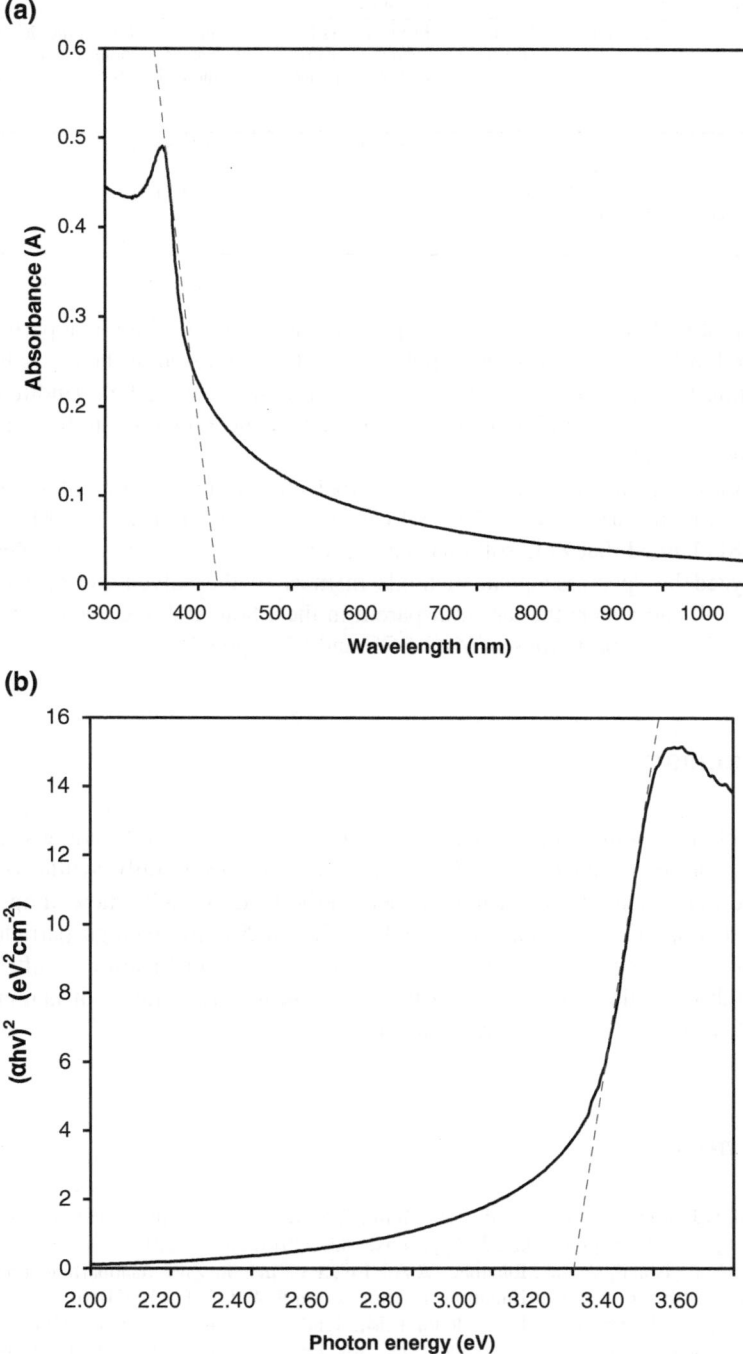

Fig. 3.12 a UV–vis absorption spectrum and **b** plot of absorption as a function of energy for the ZnO SGC powder

Table 3.1 Properties of ZnO synthesized by SGS and SGC

Properties of ZnO	Structure	Crystallite size (nm)	FESEM average particles size (nm)	TEM average particles size (nm)	Atomic percentage of zinc and oxygen (%)	Stoichiometric composition ratio	Band gap energy (E_g)
Sol–gel storage (SGS)	Hexagonal wurtzite	56	330–530		56.12:43.88	1:1	–
Sol gel centrifugation (SGC)	Hexagonal wurtzite	58	20–80	55	58.26:41.74	1:1	3.20 eV

The plot of $(\alpha h v)^2$ as a function of photon energy of the ZnO nanoparticles is illustrated in Fig. 3.12b. After extrapolating the linear portion of the curve to zero absorption, the E_g of ZnO was found to be approximately 3.20 eV (indirect E_g). This result is almost similar to those obtained in the previous studies on ZnO nanocrystalline [15, 16].

The most relevant parameter for the optoelectronic applications of ZnO is its large E_g, which varies from 3.2 to 3.4 eV at room temperature [17]. Monticone et al. [18] claimed that ZnO solid has an ~ 3 eV band gap energy. Thus, ZnO can be analyzed by spectroscopical methods in most of the solvents (e.g., CH_3OH, propan-2-ol, and water) that are transparent in the whole UV–vis spectral regions. Table 3.1 lists the properties of ZnO SGS and SGC powders.

3.6 Summary

The effect of the different procedures in the sol–gel process used to synthesize ZnO powders was investigated. The ZnO powders were successfully synthesized by SGS and SGC methods. The ZnO powder synthesized via SGC showed optimum structural properties with largest crystallite size of 58 nm, average particle size between 20 and 80 nm, and indirect E_g of 3.20 eV. Centrifugation resulted in a considerably smaller particle size of the ZnO powder in the range of nanometers and higher crystallinity compared with SGS.

References

1. Duval, D.J., McCoy, B.J., Risbud, S.H., Munir, Z.A.: Size selected silicon particles in sol–gel glass by centrifugal processing. J. Appl. Phys. **83**, 2301–2307 (1998)
2. Alias, S.S., Ismail, A.B., Mohamad, A.A.: Effect of pH on ZnO nanoparticle properties synthesized by sol–gel centrifugation. J. Alloy. Compd. **499**, 231–237 (2010)
3. Rani, S., Suri, P., Shishodia, P.K., Mehra, R.M.: Synthesis of nanocrystalline ZnO powder via sol–gel route for dye-sensitized solar cells. Sol. Energy Mater. Sol. Cells **92**, 1639–1645 (2008)

4. Li Jian, K.: Synthesis and Characterization of Zinc Oxide Powder via Sol–Gel Route for Quasi-Solid-State Solar Cells, Final Year Project Report, Universiti Sains Malaysia, Nibong Tebal, Penang, Malaysia (2009)
5. Tan, A.L., Khoo, L.J., Alias, S.S., Mohamad, A.A.: ZnO nanoparticles and poly(acrylic) acid-based polymer gel electrolyte for photo electrochemical cell. J. Sol–Gel. Sci. Technol. **64**, 184–192 (2012)
6. Maensiri, S., Laokul, P., Promarak, V.: Synthesis and optical properties of nanocrystalline ZnO powders by a simple method using zinc acetate dihydrate and poly(vinyl pyrrolidone). J. Cryst. Growth **289**, 102–106 (2006)
7. Du, H., Yuan, F., Huang, S., Li, J., Zhu, Y.: A new reaction to ZnO nanoparticles. Chem. Lett. **33**, 770–771 (2004)
8. Mazloumi, M., Taghavi, S., Arami, H., Zanganeh, S., Kajbafvala, A., Shayegh, M.R., Sadrnezhaad, S.K.: Self-assembly of ZnO nanoparticles and subsequent formation of hollow microspheres. J. Alloy. Compd. **468**, 303–307 (2009)
9. Hosono, E., Fujihara, S., Kimura, T., Imai, H.: Non-basic solution routes to prepare ZnO nanoparticles. J. Sol–Gel. Sci. Technol. **29**, 71–79 (2004)
10. Jang, J.-M., Kim, S.-D., Choi, H.-M., Kim, J.-Y., Jung, W.-G.: Morphology change of self-assembled ZnO 3D nanostructures with different pH in the simple hydrothermal process. Mater. Chem. Phys. **113**, 389–394 (2009)
11. Lin, C.-C., Li, Y.-Y.: Synthesis of ZnO nanowires by thermal decomposition of zinc acetate dihydrate. Mater. Chem. Phys. **113**, 334–337 (2009)
12. Musić, S., Dragčević, Ð., Popović, S., Ivanda, M.: Precipitation of ZnO particles and their properties. Mater. Lett. **59**, 2388–2393 (2005)
13. Elkhidir, S.A., Tang, Y., Xu, L.: Preparation of ZnO nanoparticles and nanosheets and their application to dye-sensitized solar cells. Sol. Energy Mater. Sol. Cells **91**, 1658–1662 (2007)
14. Kamalasanan, M.N., Chandra, S.: Sol-gel synthesis of ZnO thin films. Thin Solid Films **288**, 112–115 (1996)
15. Bacaksiz, E., Parlak, M., Tomakin, M., Özçelik, A., Karakız, M., Altunbaş, M.: The effects of zinc nitrate, zinc acetate and zinc chloride precursors on investigation of structural and optical properties of ZnO thin films. J. Alloy. Compd. **466**, 447–450 (2008)
16. Marotti, R.E., Guerra, D.N., Bello, C., Machado, G., Dalchiele, E.A.: Bandgap energy tuning of electrochemically grown ZnO thin films by thickness and electrodeposition potential. Sol. Energy Mater. Sol. Cells **82**, 85–103 (2004)
17. Bhargava, R.N. (ed.): Properties of Wide Bandgap II–VI Semiconductors., EMIS Datareviews Series No.17, INSPEC, London, United Kingdom.27 (1997)
18. Monticone, S., Tufeu, R.: Kanaev: complex nature of the UV and visible fluorescence of colloidal ZnO nanoparticles. J. Phys. Chem. B **102**, 2854–2862 (1998)

Chapter 4
ZnO: Photoelectrochemical Cells Analysis

Abstract The photoelectrochemical cells (PEC) basic concept and its construction is presented in this chapter. Two types of PEC are chosen as it were fabricated based on the optimal properties of ZnO synthesized by sol–gel process. Both of PEC performances are explained by the current density–voltage (J–V) properties of the PEC under illuminated and dark conditions.

Keywords Photoelectrochemical cells · Current density–voltage · Illuminated · Dark

4.1 PEC: Construction and Operation

Various types of PEC have been discovered since Fujishima and Honda [1] introduced the semiconductor-liquid electrolyte junction. Practical advances are gained annually by replacing the type of nanostructured working electrode, electrolyte, and counter electrode to increase PEC performance.

A PEC generally consists of a transparent conducting oxide-coated glass as substrate for the semiconductor working photoelectrode, a dye, a redox electrolyte, and a counter electrode. The nanocrystalline PEC is similar to natural photosynthesis in two aspects. First, the functions of the organic dye as light absorber and the production of electron flow were similar to the chlorophyll layer inside a leaf. Second, the PEC also uses multiple layers, similar to a leaf, to enhance both light absorption and electron collection efficiency [2].

A solution containing nanometer-sized particles of TiO_2 was distributed uniformly on a glass plate coated with a thin conductive and transparent layer of tin dioxide (SnO_2) to create nanocrystalline solar cell. Then, the TiO_2 film was dried and heated to form a porous and high-surface area TiO_2 structure [3].

The TiO_2 film on the glass plate was then dipped into a dye solution, such as a red ruthenium (Ru) containing organic dye or green chlorophyll derivative. In the meantime, a drop of liquid electrolyte with iodide was placed on the film to penetrate into the pores of the membrane to complete the device. A counter

S. S. Alias and A. A. Mohamad, *Synthesis of Zinc Oxide by Sol–Gel Method for Photoelectrochemical Cells*, SpringerBriefs in Materials, DOI: 10.1007/978-981-4560-77-1_4, © The Author(s) 2014

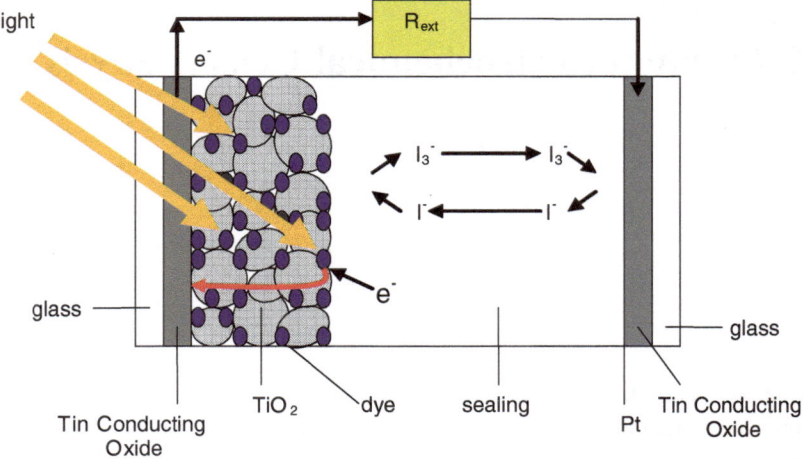

Fig. 4.1 Schematic structure and reaction in PEC

electrode of conductive glass coated with a thin catalytic layer of platinum or carbon was placed on top, and the sandwich was illuminated through the TiO_2 side [2].

In the nanocrystalline PEC, each dye layer did not absorb much light. However, the interconnected particles of the porous membrane with dye absorbed 90 % of the visible light. Almost all of the excited electrons produced from light absorption could be injected into the TiO_2 to produce electricity because the dye layer was very thin. The electrons lost by the dye via light absorption were quickly replaced by the mediator [2].

Typically, iodide ion in the electrolyte solution acted as mediator in the PEC. The oxidized mediator formed iodide/triiodide (I^-/I_3^-), which in turn obtained an electron at the counter electrode after the electron flowed through the electrical load. This reaction occurred repeatedly while the PEC was exposed to light. The reaction of produced energy in the nanocrystalline PEC is illustrated in Fig. 4.1.

4.2 ZnO: Photoelectrochemical Cell Working Electrode

Several semiconductor materials, including single-crystal and polycrystal forms of n-and p-Si, CdS–CdTe, n-and p-GaAs, n-and p-InP, and n-CdS, were used as photoelectrodes. These materials can perform well when used with a suitable redox electrolyte [4].

The oxide semiconductor materials also exhibit good stability under irradiation in solution [4]. However, stable oxide semiconductor types could not absorb

Table 4.1 Photovoltaic performance of various PEC oxide semiconductors [6]

Oxide semiconductor	J_{SC} (mA cm^{-2})	V_{OC} (V)
In_2O_3	2.38	0.39
TiO_2	2.10	0.78
ZrO_2	0.004	0.14
SnO_2	1.95	0.58
ZnO	7.44	0.52

Table 4.2 Performance of ZnO PEC working electrode [9]

Working electrode	J_{SC} (mA cm^{-2})	V_{OC} (V)	References
ZnO	1.30	0.56	[10]
ZnO	10.90	0.58	[11]
ZnO	1.50	0.60	[12]
ZnO	9.06	0.58	[13]
ZnO	6.49	0.79	[14]

visible light because the semiconductors had wide band gaps. To remedy this problem, the wide band gap oxide semiconductor materials were sensitized. The sensitization of TiO_2, ZnO (ZnO), and SnO_2 with photosensitizers, such as organic dyes, can enhance the potential of light absorption by exciting and injecting electrons into the CB of the semiconductor electrodes.

The combination of two types of oxide semiconductor photoelectrodes can also increase PEC efficiency. Tennakone et al. [5] used a porous film with a mixture of SnO_2 and ZnO sensitized using a Ru bipyridyl complex that showed good PEC performance. Table 4.1 tabulates the photovoltaic performance of the PECs using various kinds of semiconductor working electrodes [6].

The nanostructured ZnO is an attractive material that has been used extensively for optoelectronic devices, particularly for the PEC working electrode because of its wide-band gap semiconductor with good carrier mobility, and it can be doped as both n- and p-types [7]. The electron mobility is much higher in ZnO than in TiO_2, and the CB edges of both materials are located at approximately the same level [8]. The performance of the ZnO PEC working electrode is tabulated in Table 4.2. Comparisons were made based on the similarities of the types of working electrodes only. PECs had different types of electrolytes and counter electrodes.

4.3 Performance of ITO–ZnO/Gel Polymer Electrolyte (GPE)/Au–ITO Cell

The fabrication of ITO-ZnO (N719 dye)/GPE/Au–Pd-ITOPEC can be divided into three preparation parts. The ITO-glass substrates with 8 Ω cm^{-2} sheet resistance were cleaned using acetone in an ultrasonic bath. One of the ITO glass coated with

ZnO paste (from Chap. 2) acted as the working electrode. Another ITO glass was employed as counter electrode by coating with Au–Pd.

For the working electrode, the ZnO paste synthesized at pH 9 (Sect. 2.1) was coated on the ITO-glass layer. The ZnO powder (0.5 g) synthesized in Sect. 2.1 was mixed with another two materials. First, dilute acetic acid (0.035 M CH_3COOH) was added to form a slightly viscous colloidal suspension. Second, a few drops of Triton X-100 $[C_{14}H_{22} (C_2H_4O)N]$ were added to bind the suspension. All these materials were ground together.

The ZnO paste was then spread on an ITO glass using a glass rod. Initially, adhesive tape was used as spacer, leaving an area of 1 cm^2 on the ITO glass for the ZnO paste. Then, the ZnO paste layer on the ITO-glass was dried in air and sintered at 400 °C for 30 min in Lenton EHF 1,700. The electrode was left to cool at 25 °C.

The ZnO electrode was dipped into N-719 dye for 1 h. The 0.5 mM N-719 dye was diluted into 100 ml of anhydrous ethanol C_2H_5OH. The ZnO electrode was dried in air before further fabrication as a PEC.

To fabricate the PEC, the GPE with the highest room-temperature conductivity was sandwiched by the working and counter electrodes. The silver paste was coated at the edge of both working and counter electrodes for better connection of the PEC electrodes during testing. The preparation process of the PEC components is depicted in Fig. 4.2.

The ITO–ZnO (N719 dye)/GPE/Au–Pd–ITOPEC was then characterized by its performance by measuring in triplicate the J–V, short circuit current density (J_{SC}), and output circuit voltage (V_{OC}) under light and dark conditions. The maximum current density (J_{max}) and maximum voltage (V_{max}) were obtained from the area under J–V curve. The J–V characteristics for biased PEC were obtained by connecting the cell to Autolab PGSTAT 30 GPES (Eco Chemie, B.V.). The setup components are shown in Fig. 4.3.

The values of J_{SC} was obtained from the highest value of Y-intercept at x = 0, while V_{OC} value was from the highest value of X-intercept at y = 0. J_{max} and V_{max} were obtained from a square shaped area under the J–V curve. The fill factor (FF) and overall conversion efficiency percentage (η %) were calculated using Eqs. 4.1 and 4.2:

$$FF = \frac{(V_{max} \times J_{max})}{(V_{oc} \times J_{sc})} \tag{4.1}$$

$$\eta(\%) = \left[\frac{(V_{max} \times J_{max})}{P_{in}}\right] \times 100\%$$
$$= \left[\frac{(V_{oc} \times J_{SC} \times FF)}{P_{in}}\right] \times 100\% \tag{4.2}$$

where P_{in} is the incident light power (100 mW cm^{-2}).

Figure 4.4 depicts the J–V curve of the PEC-based ITO–ZnO at pH 9 (N719 dye)/GPE/Au–Pd–ITO under dark and illuminated (100 mW cm^{-2}) conditions.

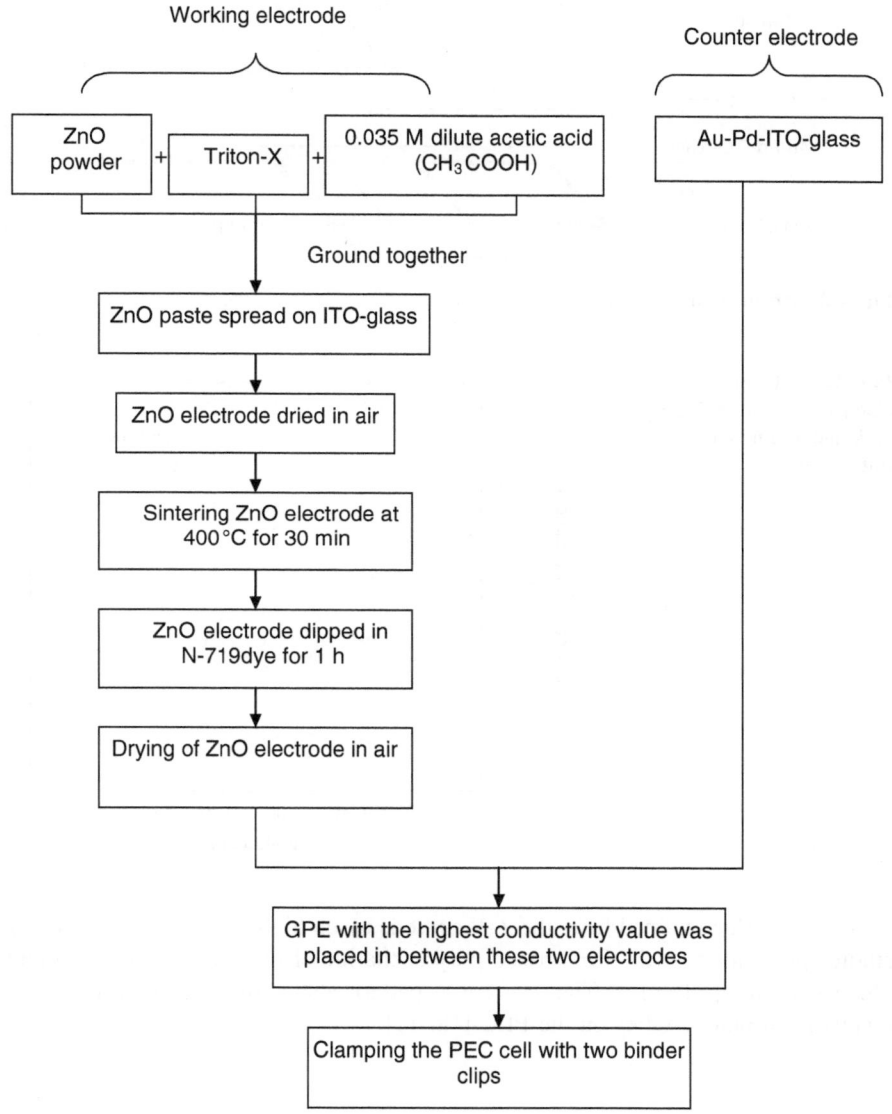

Fig. 4.2 Preparation of PEC components [15]

The J–V curve is much lower in the dark condition compared with the curve under illumination. However, when the PEC was exposed under 100 mW cm^{-2} of halogen lamp, the V_{OC} and J_{SC} values were 0.29 V and 0.007 mA cm^{-2}, respectively. The V_{max} and J_{max} values obtained in this part were 0.17 V and 0.005 mA cm^{-2}, respectively. The FF and η values were 0.41 and 0.008 %.

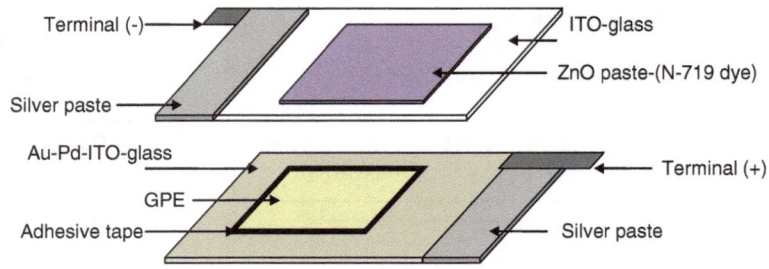

Fig. 4.3 Schematic diagram of PEC components

Fig. 4.4 *J–V* curve characteristic of PEC under dark and 100 mW cm^{-2} illumination

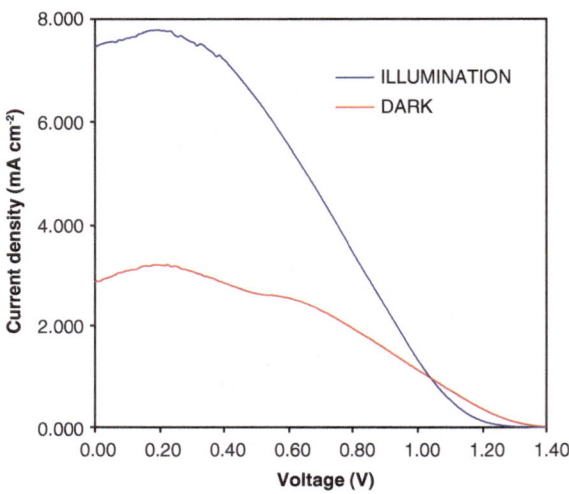

The J_{SC} value attained from the *J–V* curve is much higher compared with other studies performed on the PEC-based polysaccharide electrolyte [from 2 to 5 and 19.23 µA cm^{-2}] [16, 17]. The *J–V* curve pattern also shows a similarity to the results from other studies on the PEC [18, 19].

4.4 Performance of Cu–ZnO/GPE/C–ITO Cell

The selection of ZnO powder from the SGC and SGS procedures in sol–gel route in Sect. 3.1 to fabricate solar cells was based on the XRD and FESEM results. The flow chart for the general process flow in the fabrication and characterization of solar cells is depicted in Fig. 4.5.

The ZnO paste for solar cell fabrication was prepared by grinding approximately 0.5 g of nanocrystalline ZnO powder (synthesized by SGC method) in a mortar and pestle. A few drops of highly diluted acetic acid (0.035 M), which was

Fig. 4.5 Flow chart showing
the general process flow for
the fabrication and
characterization of solar cells
[20]

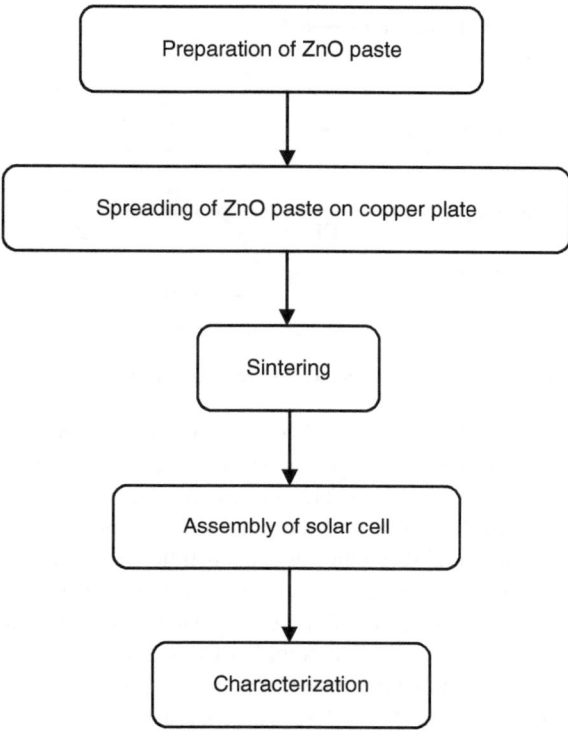

prepared by adding 0.1 ml of concentrated acetic acid to 50 ml of water, were
added to the powder. The grinding and addition of highly diluted acetic acid were
performed alternatively to form a slightly soupy colloidal suspension with smooth
consistency. Afterward, a few drops of Triton X-100 were added as binder to the
suspension.

Prior to the spreading of the ZnO paste, a copper plate was ground to remove
the oxide layer and then cleaned using ethanol in an ultrasonic cleaning machine
by Sono Swiss. The cleaned copper plate was taped on four sides using one
thickness of scotch tape leaving an area of 1 cm², and any oils or fingerprints were
wiped off using a wet tissue with ethanol. Afterward, the prepared ZnO paste was
spread on the copper plate using a glass rod. The thickness of the ZnO paste layer
on the copper substrate was controlled by the tape which acted as a 40 μm to
50 μm spacer. Afterward, the tapes were removed carefully without scratching the
ZnO paste layer.

Next, the ZnO paste layer on the copper substrate was dried in air and sintered
at 400 °C for 30 min in a furnace by carbolite. This sintering step heated the
sample to dry and burnt the organic solvent and surfactant so that ZnO paste would
have good contact to the copper substrate. The surface of the sample turned grey
after sintering.

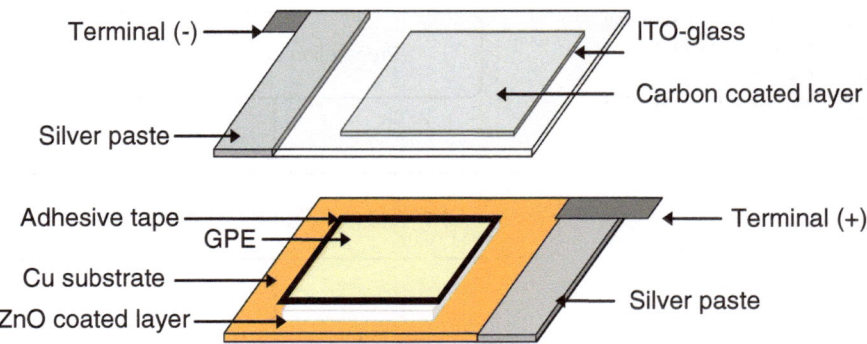

Fig. 4.6 Schematic structure and reaction in PEC

The solar cell was fabricated by sandwiching the GPE between the ITO glass and ZnO on the copper substrate. The GPE used was produced according to the study by Tan et al. [21]. This electrolyte consisted of 0.8 M sodium iodide (NaI) and 0.02 M iodine (I_2) trapped in 0.5 g poly(acrylic acid) (PAA).

After the ZnO on the copper substrate was sintered, double-sided tapes were put at four sides of the ZnO layer, and GPE (PAA: NaI + I_2) was inserted between the double-sided tapes. This tape served to prevent the leakage of the electrolyte from the solar cell. Prior to the assembly of Cu–ZnO/PAA: NaI + I_2 with ITO, the ITO glass was cleaned using ethanol in an ultrasonic cleaning machine. With the conducting side down, the ITO glass was coated with carbon. Afterward, the solar cell was fabricated by placing carbon-coated ITO glass on Cu–ZnO/PAA: NaI + I_2 with the carbon-coated side facing Cu–ZnO/PAA: NaI + I_2, and then clamped together by two binder clips. The carbon-coated ITO glass and Cu–ZnO/PAA: NaI + I_2 plate were offset so that the uncoated glass extended beyond the sandwich. The layout of fabricated Cu–ZnO/PAA: NaI + I_2/ITO solar cell is demonstrated in Fig. 4.6.

Fig. 4.7 *J–V* characteristic of Cu–ZnO/PAA + 0.8 M NaI + 0.02 M I_2/C–ITO cell in the dark and under illumination

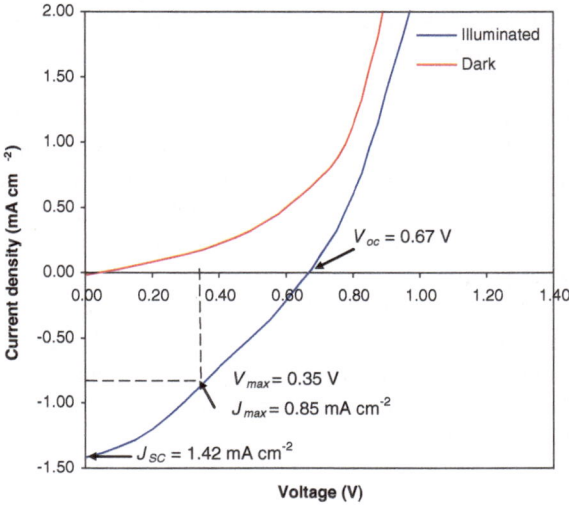

The J–V curves of Cu–ZnO/P0.8N0.02I PGE/C-ITO PEC are demonstrated in Fig. 4.7. The V_{OC} and J_{SC} were found to be 0.67 V and 1.42 mA cm^{-2}, respectively. According to prior calculations, V_{max} (0.35 V) and J_{max} (0.85 mA) were determined from the point of maximum power on the J–V curve. Then, FF (0.312) and η (0.30 %) of the PEC were calculated from Eqs. (4.1) and (4.2). The V_{OC} of the photochemical cell was in the range of the electrochemical stability window determined from the linear sweep voltammetry (LSV) results (1.60 V), suggesting that the PGE is applicable in the PEC electrolyte, similar to previous studies [22–24].

4.5 Summary

ZnO has been widely used for optoelectronic devices mainly as a PEC working electrode because of its wide-band gap semiconductor with good carrier mobility, and its capability to be doped both as n- and p-type. The optimal properties of the ZnO nanoparticles synthesized at pH 9 and SGC have been fabricated as ITO-ZnO/GPE/Au-ITO and Cu–ZnO/GPE/C–ITO cell. The results show that the ZnO electrode is a good semiconductor material with good stability for PEC working electrode application.

References

1. Fujishima, A., Honda, K.: Electrochemical photolysis of water at a semiconductor electrode. Nature **238**, 37–38 (1972)
2. Smestad, G.P.: Education and solar conversion: demonstrating electron transfer. Sol. Energy Mater. Sol. Cells **55**, 157–178 (1998)
3. Li, B., Wang, L., Kang, B., Wang, P., Qiu, Y.: Review of recent progress in solid-state dye-sensitized solar cells. Sol. Energy Mater. Sol. Cells **90**, 549–573 (2006)
4. Hara, K., Arakawa, H.: Dye-sensitized solar cells. In: Handbook of Photovoltaic Science and Engineering, pp. 663–700. Wiley, USA (2005)
5. Tennakone, K., Kumara, G.R.R.A., Kottegoda, I.R.M., Perera, V.P.S.: An efficient dye-sensitized photoelectrochemical solar cell made from oxides of tin and zinc. Chem. Commun. 15–16 (1999)
6. Hara, K., Horiguchi, T., Kinoshita, T., Sayama, K., Sugihara, H., Arakawa, H.: Highly efficient photon-to-electron conversion with mercurochrome-sensitized nanoporous oxide semiconductor solar cells. Sol. Energy Mater. Sol. Cells **64**, 115–134 (2000)
7. Ip, K., Thaler, G.T., Yang, H., Youn Han, S., Li, Y., Norton, D.P., Pearton, S.J., Jang, S., Ren, F.: Contacts to ZnO. J. Cryst. Growth **287**, 149–156 (2006)
8. Boschloo, G., Edvinsson, T., Hagfeldt, A.: Chapter 8—dye-sensitized nanostructured ZnO electrodes for solar cell applications. In: Tetsuo, S. (ed.) Nanostructured Materials for Solar Energy Conversion, pp. 227–254. Elsevier, Amsterdam (2006)
9. Gonzalez-Valls, I., Lira-Cantu, M.: Vertically-aligned nanostructures of ZnO for excitonic solar cells: a review. Energy Environ. Sci. **2**, 19–34 (2009)
10. Hosono, E., Fujihara, S., Honma, I., Zhou, H.: The fabrication of an upright-standing zinc oxide nanosheet for use in dye-sensitized solar cells. Adv. Mater. **17**, 2091–2094 (2005)

11. Keis, K., Magnusson, E., Lindström, H., Lindquist, S.-E., Hagfeldt, A.: A 5% efficient photoelectrochemical solar cell based on nanostructured ZnO electrodes. Sol. Energy Mater. Sol. Cells **73**, 51–58 (2002)
12. Kakiuchi, K., Hosono, E., Fujihara, S.: Enhanced photoelectrochemical performance of ZnO electrodes sensitized with N-719. J. Photochem. Photobiol. A-Chem. **179**, 81–86 (2006)
13. Ku, C.-H., Wu, J–.J.: Chemical bath deposition of ZnO nanowire–nanoparticle composite electrodes for use in dye-sensitized solar cells. Nanotechnology **18**, 505706 (2007)
14. Goncalves, A.d.S., Davolos, M.R., Masaki, N., Yanagida, S., Morandeira, A., Durrant, J.R., Freitas, J.N., Nogueira, A.F.: Synthesis and characterization of ZnO and ZnO:Ga films and their application in dye-sensitized solar cells, Dalton Trans. 1487–1491 (2008)
15. Alias, S.S.: Synthesis and Characterization of Agar Gel Polymer Electrolyte and Zinc Oxide for Photoelectrochemical Cell. Masters of Science, Universiti Sains Malaysia, Nibong Tebal, Penang, Malaysia (2011)
16. Buraidah, M.H., Teo, L.P., Majid, S.R., Arof, A.K.: Characteristics of TiO_2/solid electrolyte junction solar cells with redox couple. Opt. Mater. **32**, 723–728 (2010)
17. Mohamad, S.A., Yahya, R., Ibrahim, Z.A., Arof, A.K.: Photovoltaic activity in a ZnTe/PEO–chitosan blend electrolyte junction. Sol. Energy Mater. Sol. Cells **91**, 1194–1198 (2007)
18. Kaneko, M., Hoshi, T.: Dye-sensitized solar cell with polysaccharide solid electrolyte. Chem. Lett. **32**, 872–873 (2003)
19. Philias, J.M., Marsan, B.: All-solid-state photoelectrochemical cell based on a polymer electrolyte containing a new transparent and highly electropositive redox couple. Electrochim. Acta **44**, 2915–2926 (1999)
20. Li Jian, K.: Synthesis and characterization of zinc oxide powder via sol-gel route for quasi-solid-state solar cells. Final Year Project Report, Universiti Sains Malaysia, Nibong Tebal, Penang, Malaysia (2009)
21. Tan, A.L., Khoo, L.J., Alias, S.S., Mohamad, A.A.: ZnO nanoparticles and poly(acrylic) acid-based polymer gel electrolyte for photo electrochemical cell. J. Sol-Gel. Sci. Technol. **64**, 184–192 (2012)
22. Guillén, E., Fernández-Lorenzo, C., Alcántara, R., Martín-Calleja, J., Anta, J.A.: Solvent-free ZnO dye-sensitised solar cells. Sol. Energy Mater. Sol. Cells **93**, 1846–1852 (2009)
23. Stergiopoulos, T., Arabatzis, I.M., Cachet, H., Falaras, P.: Photoelectrochemistry at SnO_2 particulate fractal electrodes sensitized by a ruthenium complex: Solid-state solar cell assembling by incorporating a composite polymer electrolyte. J. Photochem. Photobiol. A-Chem. **155**, 163–170 (2003)
24. Tan, W.C., Alias, S.S., Ismail, A.B., Mohamad, A.A.: Effect of styrene–acrylonitrile content on 0.5 M NaI/0.05 M I_2 liquid electrolyte encapsulation for dye-sensitized solar cells. J. Solid State Chem. **16**, 2103–2112 (2012)

Index

S. S. Alias and A. A. Mohamad, *Synthesis of Zinc Oxide by Sol–Gel Method*   51
for Photoelectrochemical Cells, SpringerBriefs in Materials,
DOI: 10.1007/978-981-4560-77-1, © The Author(s) 2014